Exercises in Organic Synthesis Based on Synthetic Drugs

Authored by

Marcus Vinícius Nora de Souza
FioCruz-Fundação Oswaldo Cruz
Rio de Janeiro - RJ, 21040-900
Brazil

Exercises in Organic Synthesis Based on Synthetic Drugs

Author: Marcus Vinícius Nora de Souza

ISBN (Online): 978-981-14-8756-9

ISBN (Print): 978-981-14-8754-5

ISBN (Paperback): 978-981-14-8755-2

need for a court order if at any point you breach any terms of this License Agreement. In no event will any delay or failure by Bentham Science Publishers in enforcing your compliance with this License Agreement constitute a waiver of any of its rights.

3. You acknowledge that you have read this License Agreement, and agree to be bound by its terms and conditions. To the extent that any other terms and conditions presented on any website of Bentham Science Publishers conflict with, or are inconsistent with, the terms and conditions set out in this License Agreement, you acknowledge that the terms and conditions set out in this License Agreement shall prevail.

Bentham Science Publishers Pte. Ltd.
80 Robinson Road #02-00
Singapore 068898
Singapore
Email: subscriptions@benthamscience.net

CONTENTS

PREFACE

Chemistry is a science based on compositions, properties, reactions and transformations of substances. In 1777, it was divided into two main fields by the Swedish chemist Torbern Olof Bergman: organic and inorganic chemistry. However, due to the vast amount of information obtained over the centuries, modern chemistry has been divided into five main branches: organic, inorganic, analytical and physical chemistry, and biochemistry. Nowadays, despite this division, these main branches are subdivided into several sub-branches demonstrating the tremendous scientific and technological development of chemistry. In the case of organic chemistry, it presented a large number of sub-branches, such as organic synthesis, medicinal chemistry, asymmetric synthesis, natural products, green chemistry, organometallic chemistry, physical organic chemistry, polymer chemistry, forensic chemistry, among others. Due to the importance of chemistry as the physical sciences in our world, many researchers considered it as a central science closely linked with applied and life sciences.

Chemistry plays a critical role in drug discovery against the most different types of diseases. The development of a new drug is an extremely complex process involving different areas and, types of knowledge. Considering that, two sub-branches of organic chemistry are fundamental at the beginning of this process: organic synthesis and medicinal chemistry. Organic synthesis is the preparation in the laboratory of simple or complex substances using chemical reactions in a rational and planned way, starting with simpler atoms or elements. Medicinal chemistry is a discipline that is at the interface of several areas, such as, chemistry, medical, biological, and pharmaceutical sciences. Its objective is the design, discovery, identification, and preparation of bioactive compounds as well as the study of other aspects such as their mechanisms of action, metabolism, and a structure-activity relationship.

Due to the crucial importance of organic synthesis and medicinal chemistry in the development of new drugs, the objective of this book is to present different exercises aiming at the search for biologically active compounds against the most diverse types of diseases in both humans and animals. The development of a new drug begins with the search, planning, design, and evaluation being necessary for its preparation in the laboratory. This initial step is a long and complicated process for a biologically active compound to become a drug. For this to happen, organic synthesis is of fundamental importance since, through it, we can obtain the most varied types of simple or complex substances for biological evaluation. Thus, the present book will address different techniques and strategies for the synthesis of biologically active compounds and drugs. Through the exercises presented, the reader will be able to propose, obtain, and evaluate in a logical way compounds to be studied. Considering that, this book is useful for students of graduation and post-graduation in **chemistry**.

It is worth mentioning that the present book has a wide range of bibliographical references, extracted from different scientific journals and books. The exercises presented are based on this material. In case the reader wants to delve into specific content, drug, or illness, it is advisable to consult such references. This book will be studied only for the drugs and biologically active compounds developed in the laboratory.

CONSENT FOR PUBLICATION

Not applicable

ii

CONFLICT OF INTEREST

The author confirms that there is no conflict of interest.

ACKNOWLEDGEMENTS

Declared none.

Marcus Vinícius Nora de Souza
FioCruz-Fundação Oswaldo Cruz
Rio de Janeiro - RJ, 21040-900
Brazil
E-mail: mvndesouza@gmail.com

CHAPTER 1

How to Build a Drug: Exercises Based on the Concept of Retrosynthetic Analysis

Abstract: Concepts involving retrosynthesis were and are essential to the development of organic synthesis as science and art. Therefore, the first four exercises are based on this concept. These exercises were subdivided into themes and degrees of increasing difficulty so that the reader could gradually progress and become familiar with their concepts. Thus, the first exercise begins with the retrosynthetic analysis of drugs based on the identification of only two synthons and their respective synthetic equivalents (reagents and substances) necessary for their synthesis. Exercises **2**, **3**, and **4** are thematic and based on the identification of epoxides, piperazines, and aldehydes, used in the synthesis of drugs and bioactive substances as well as other synthons and their respective synthetic equivalents present in the exercises.

Keywords: Aldehydes, Drugs, Epoxides, Exercises, Medicinal chemistry, Organic synthesis, Piperazines, Reagentes, Retrosynthetic analysis, Substances, Synthons.

INTRODUCTION

For a new drug to reach the pharmacy shelf, a long and complicated process is necessary also involving large sums of money. This process requires multiple steps with a large number of specificities and areas such as chemists, physicians, pharmacologists, biochemists, economists, lawyers, among others. In this context, one of the first steps in this process is to obtain biologically active compounds in laboratories for evaluation, and for this, the use of organic synthesis is critical.

It is no exaggeration to mention that at present modern organic synthesis can synthesize in logic and planned manner any substance of natural or synthetic origin thanks to the contribution of several scientists in various periods of human history. In modern organic chemistry, professor Robert Bruns Woodward (1917-1979) (Nobel Prize in 1965) is considered by many as the father of modern organic synthesis for his outstanding contributions in the synthesis of natural products with extremely complex chemical structures (Fig. **1**). Another important scientist is professor Elias James Corey (1928-) (Nobel Prize in 1990), who introduced and successfully applied in the synthesis of natural products the concept of retrosynthetic analysis. This concept can efficiently and rationally simplify the synthesis of the target molecule with simple or complex chemical structures by retroactive analysis of the substance in question up to the starting materials. The target molecule is logically disconnected in parts until it reaches the starting materials and these disconnections, named by Corey as synthons, which are fragments with nucleophilic and electrophilic centers being synthetic equivalents (reagents or building blocks). In 1989, the book, *The Logic of Chemical Synthesis* brilliantly demonstrated the application of his concepts (Fig. 2).

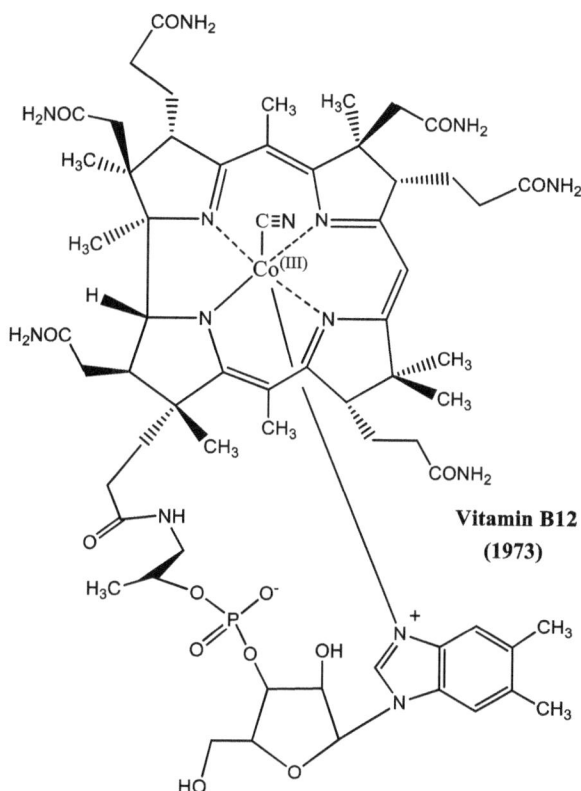

Fig. (1). Professor Robert Burns Woodward (1917-1979) and the chemical structure of vitamin B12, which to date has been the first and only synthesis to be performed in the laboratory of this complex natural product. This synthesis, which has more than one hundred steps, was carried out in collaboration with the Swiss Albert Eschenmoser (1925-) in the early 1960s with a team of approximately one hundred people and was published in 1973.
Source of the picture: The Nobel prize website.
https://www.nobelprize.org/prizes/chemistry/1965/woodward/facts/

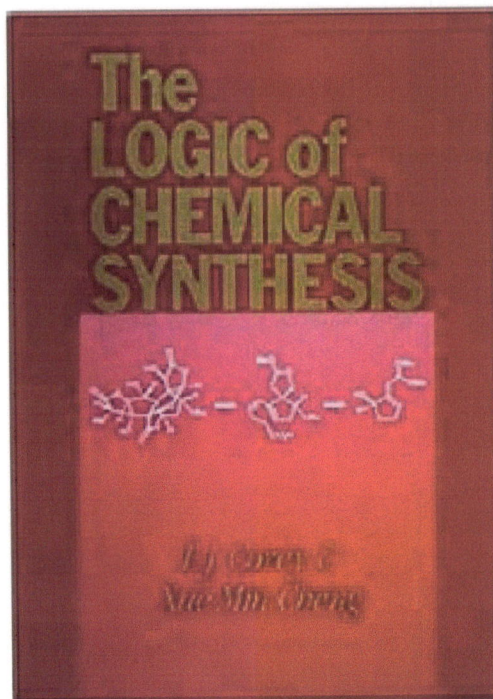

Fig. (2). Professor Elias James Corey (1928-) and his famous book.
Source of the picture: The Nobel prize website.
https://www.nobelprize.org/prizes/chemistry/1990/corey/biographical/

As an example of the application of the concepts developed by Professor Corey, we will propose the synthesis of the drug nevirapine (Fig. **3**). This drug is an anti-HIV drug approved in 1996 by the FDA (Food and Drug Administration), being a non-nucleoside reverse transcriptase inhibitor (NNRTI). At first glance, analyzing its chemical structure seems to be a very complicated task to propose its synthesis. However, when we use the retrosynthetic analysis, disconnecting our target molecule (nevirapine), we created logically nucleophilic and electrophilic centers (synthons). As we know, amines are nucleophilic and carbonyl groups reacting with nucleophiles due to their electrophilic character indicating which reagents will be used in this synthesis. It is essential to highlight how the retrosynthetic analysis was able to simplify the synthesis of the anti-HIV nevirapine. This concept is highly recommended in future synthetic analyzes.

?

How to synthesized

Nevirapine
Target molecule

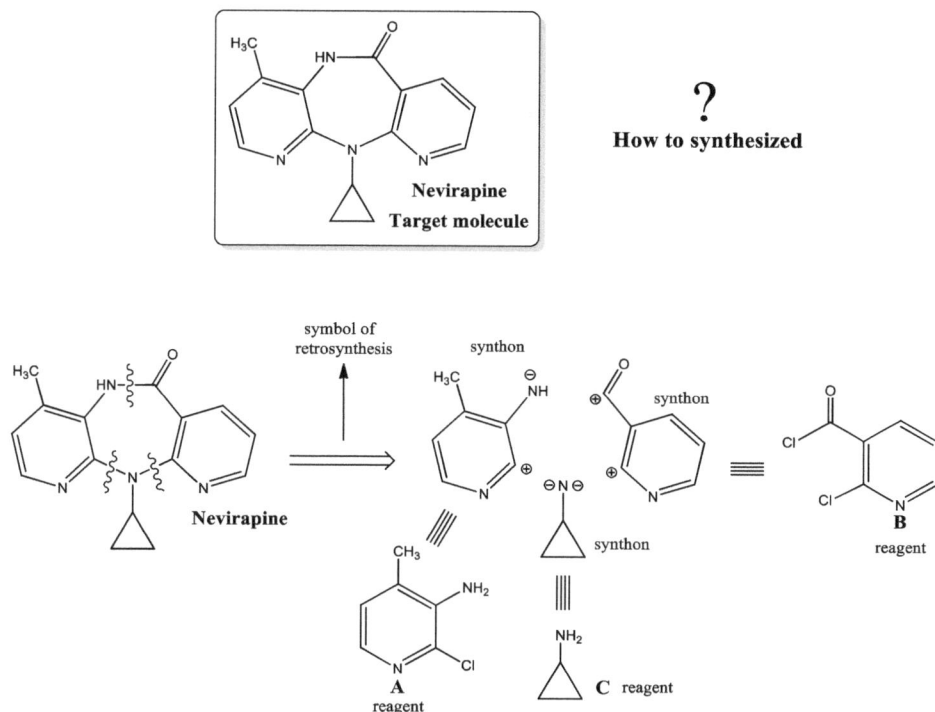

Fig. (3). Retrosynthetic analysis of the anti-HIV drug nevirapine.

PROTECTIVE GROUPS

Protective groups are a fundamental topic in organic synthesis. Without them, the organic synthesis would not have reached a high degree of scientific development in which it currently finds itself. A point to be mentioned is that, due to the importance of protective groups, it is impossible to approach organic synthesis without the protective groups being present. The definition of a protecting group in a molecule is to make a particular functional group inert. Therefore, this group would not react, and chemical transformations in this compound could be carried out. After the reactions have been made, the protecting group is removed, and the chemical function regenerated. A classic example is in Scheme **1**, in which to make a selective reduction, it is necessary to use a protective group ethylene glycol. After the transformation of the ester group, the aldehyde is regenerated by removing the protection.

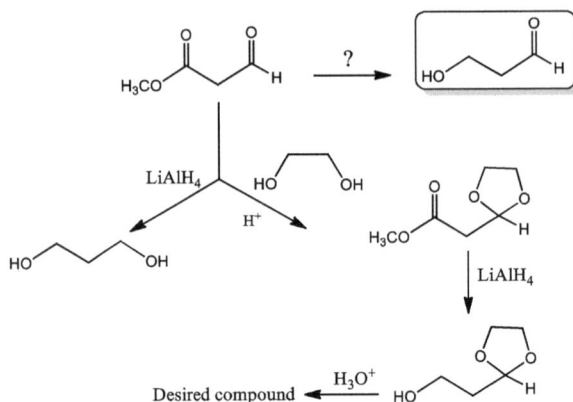

Scheme (1). Example of the concept of a protecting group in a difunctionalized molecule.

The importance of the protecting group can be seen in the synthesis of the drug sulfanilamide (4-aminobenzenesulfonamide). This drug was one of the first of this class used as a bacterial agent. The acetyl group inactivates the amino group and also hinders the reaction in position 2 in the next step. In the final stage, the protecting group was removed, and the amino group regenerates (Scheme **2**).

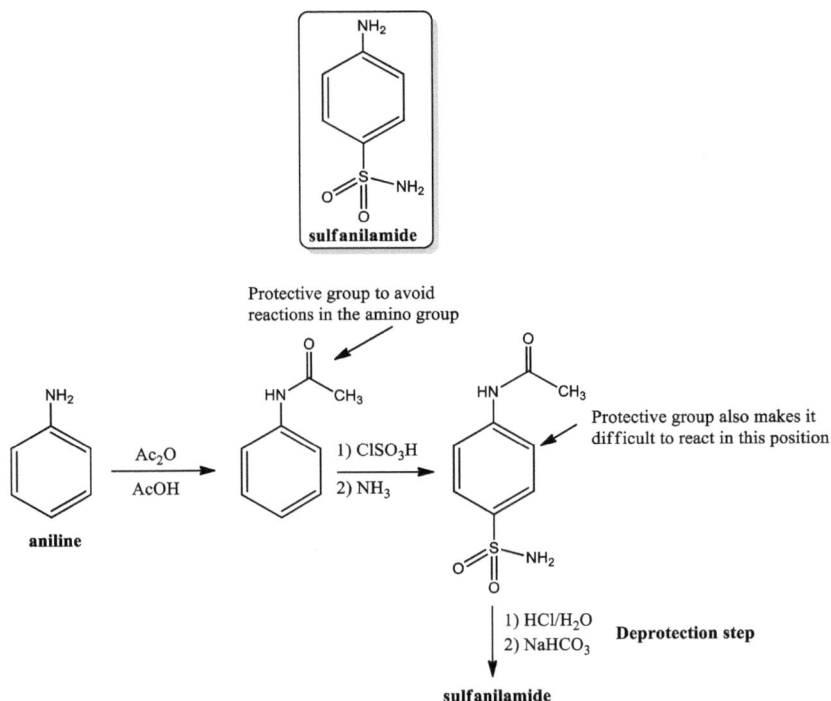

Scheme (2). Synthesis of the drug sulfanilamide.

There is a wide variety of specific protecting groups for the most diverse types of functional groups, and if the reader wants to go deeper into the topic, there are particular books on the theme [1, 2]. Due to their relevance, some of the most common protective groups will be mentioned in the exercises.

EXERCISES 1 TO 4 - RETROSYNTHETIC ANALYSIS

The first four exercises present the analysis of a given target molecule retrosynthetically. The study and planning must be performed backward, starting from the target molecule, thus identifying, through disconnections, the synthons (fragments with charges) and consequently, the substances and or reagents to be used. This type of strategy is essential in organic synthesis, since it allows, in a logical way, the choice of a specific synthetic route for the construction of the desired target molecule. It is highly recommended to use it whenever possible. Concepts involving retrosynthesis were and are essential to the development of organic synthesis as science and art. Due to the contribution of Professor Elias James Corey (1928-) in this field, he awarded the Nobel Prize in chemistry in 1990. Because of its crucial importance for the learning of organic synthesis, whenever possible, these concepts will be used throughout this book.

The first four exercises were subdivided into themes and degrees of increasing difficulty so that the reader could gradually progress and become familiar with their concepts. Thus, the first exercise begins with the retrossynthetic analysis of drugs based on the identification of only two synthons, as well as their respective synthetic equivalents (reagents and substances) necessary for their synthesis. Exercises **2**, **3** and **4** are thematic and based on the identification of epoxides, piperazines, and aldehydes, used in the synthesis of drugs and bioactive substances. The reader also has to identify the other synthons and their respective synthetic equivalents present in the exercises.

The bibliographical references mentioned in exercises **1** to **9** are in two different ways. In the first case, the drugs are in specialized books and databases that are at the end of exercise **9** [1 - 5]. The second way is to mention the bibliographic reference in a summarized way below the target molecule in question.

EXERCISE 1 - TWO SYNTHONS = DRUG

The exercises below presented several drugs used in the treatment of different diseases. Based on the concepts of retrosynthesis and synthon, the reader is invited to identify components (reagents and substances) for the preparation of these drugs. This first exercise will start the retrosynthetic analysis by the identification of only two synthons, as well as their respective synthetic equivalents (reagents and substances).

Example

Identify the two synthons employed for the synthesis of anxiolytic barbituric acid.

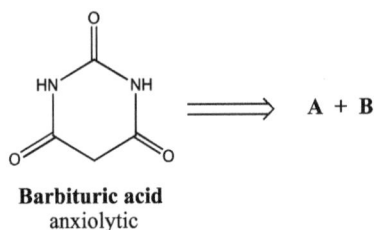

Barbituric acid
anxiolytic

Answer

When analyzing a target molecule retrospectively, it is always important to observe if it has elements of symmetry, as well as if the reader previously knows reagents or substances. This information can direct it to a good and coherent synthetic proposal. For example, the target molecule in question, a barbituric acid, we can quickly identify two known reagents after retrosynthetic analysis. One of them is urea, and the other is an α-dicarbonylated substance, for example, diethyl malonate. However, if the reader has not recognized the reagents, the analysis of the nucleophilic and electrophilic centers, can solve the problem.

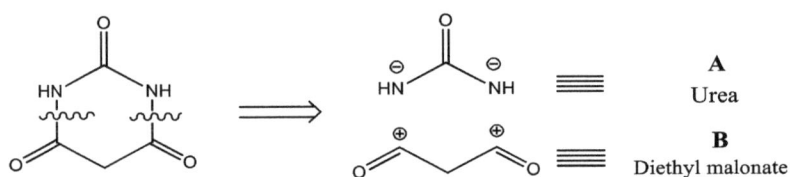

A
Urea

B
Diethyl malonate

EXERCISES

1)

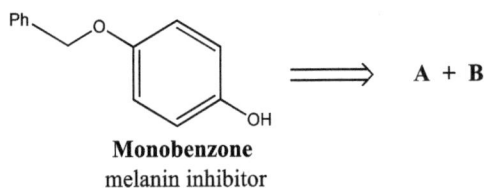

Monobenzone
melanin inhibitor

2)

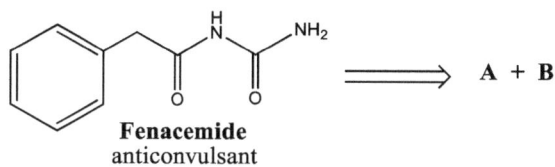

Fenacemide
anticonvulsant

$\Longrightarrow$ **A + B**

3)

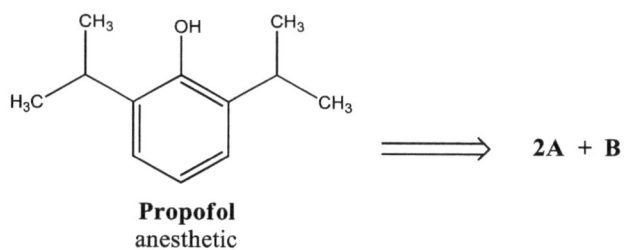

Propofol
anesthetic

$\Longrightarrow$ **2A + B**

4)

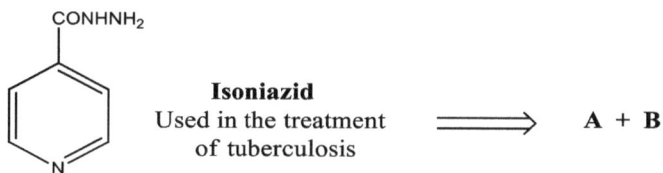

Isoniazid
Used in the treatment
of tuberculosis

$\Longrightarrow$ **A + B**

5)

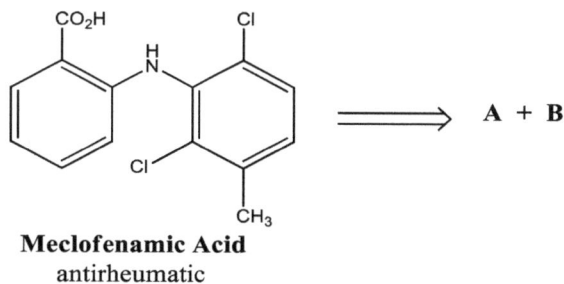

Meclofenamic Acid
antirheumatic

$\Longrightarrow$ **A + B**

6)

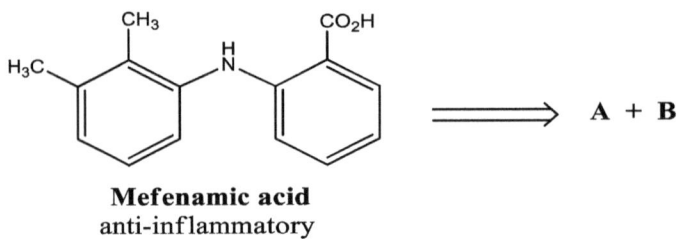

Mefenamic acid
anti-inflammatory

$\Longrightarrow$ **A + B**

7)

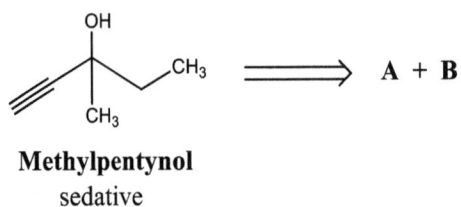

Methylpentynol
sedative

$\Longrightarrow$ **A + B**

8) In Exercises **8** and **9**, the **A** and **B** synthons are based on the formation of a cyclic substance.

Fensuximide
antiepileptic

$\Longrightarrow$ **A + B**

9)

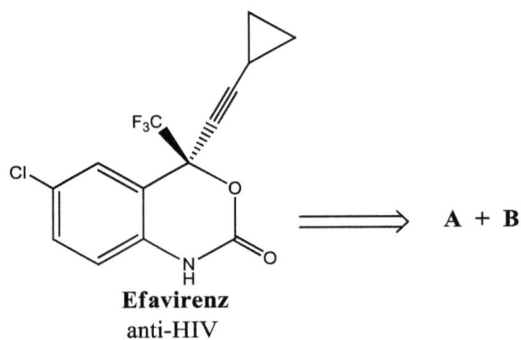

Efavirenz
anti-HIV

$\Longrightarrow$ **A + B**

10)

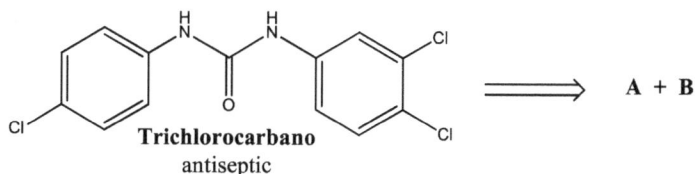

Trichlorocarbano
antiseptic

$\Longrightarrow$ **A + B**

Answers

1) In this exercise, depending on the retrosynthetic analysis and disconnection, we have two synthetic proposals. As we have already mentioned, it is always advisable to begin to identify nucleophile centers first, since they are easier to recognize. In this context, it is possible to propose two types of retrosynthetic analysis, which will consequently provide us with different synthons and reagents depending on the kind of disconnection performed. In the first proposal, we have a synthesis based on a bimolecular or second-order nucleophilic substitution reaction (SN2) between benzyl bromide and hydroquinone (1,4-dihydrox--benzene). However, in the second proposal, when we disconnect the aromatic ring, we have different synthons from the first proposal. Consequently, this provides reagents, reactions, and a synthetic route different from the first one. Thus, in this proposal, we have a nucleophilic aromatic substitution reaction between benzyl alcohol and 1-fluoro-4-hydroxybenzene. This exercise, though simple, shows us how crucial retrosynthetic analysis is and how it can provide us with different synthetic routes. These alternatives offer other alternatives if a proposed synthetic route does not work or is not as effective when tested in the laboratory.

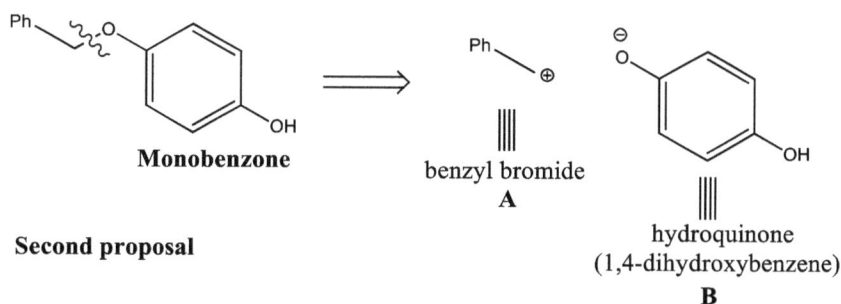

Monobenzone

Second proposal

benzyl bromide
A

hydroquinone
(1,4-dihydroxybenzene)
B

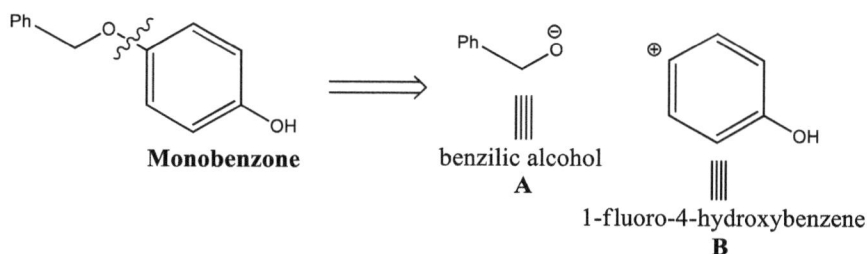

Monobenzone

benzilic alcohol
A

1-fluoro-4-hydroxybenzene
B

2) In a retrosynthetic analysis, it is always important to recognize reagents or substances that are commonly used in different syntheses, as this helps us in the disconnections of the target molecule, as well as directs us towards a sound and coherent synthetic proposal. In this exercise, we can easily recognize the substance urea, so disconnection indicates that the electrophilic center could be an acid chloride or an ester.

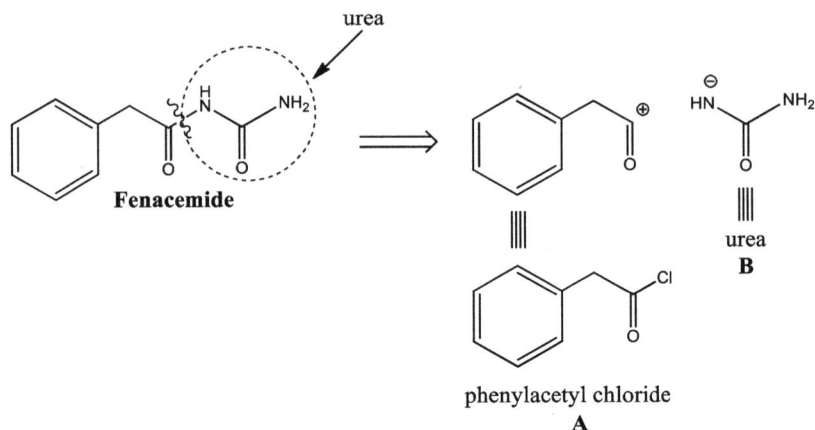

Fenacemide

phenylacetyl chloride
A

urea
B

3) A = phenol; **B** = 1-propene.

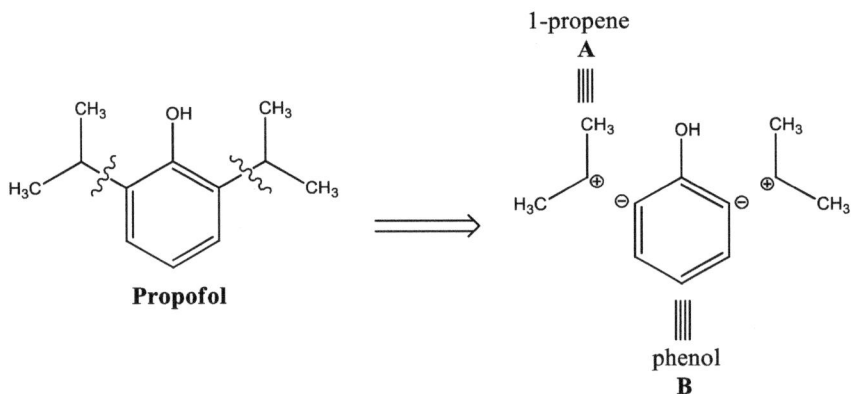

Propofol

1-propene
A

phenol
B

4)

Isoniazid   **A**   **B**

hydrazine

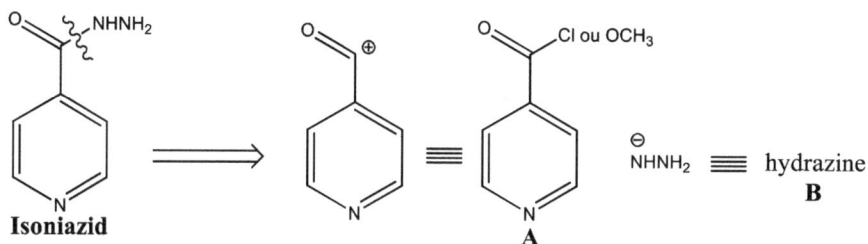

5) It is possible to perform two types of disconnections on the nucleophilic center (NH), thus obtaining two different products in each of the two proposals. The purpose of the next question is to point out whether both plans are correct or just one of them. The answer is simple since in the first proposal substance **B** (2,3,4-trichlorotoluene) can undergo nucleophilic attack in three different types of position. We are thus obtaining three different products, which would not be synthetically convenient. For this reason, the correct proposal would be the second one, since with the use of substance **B** (2,6-dichloro-3-methylaniline) we will obtain a single product, that is, the desired drug.

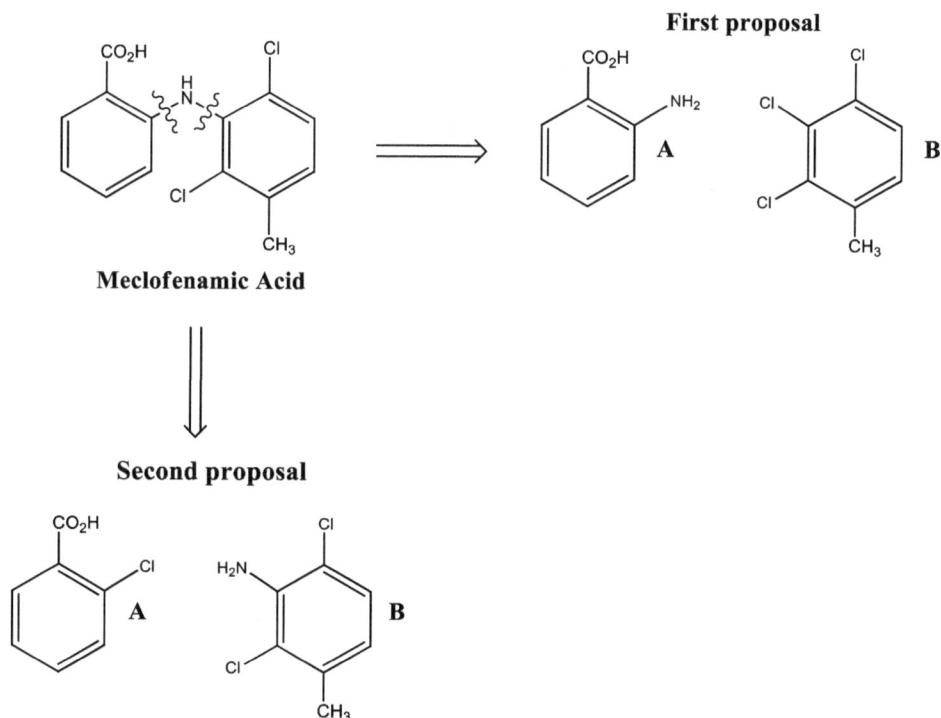

First proposal

Meclofenamic Acid   **A**   **B**

Second proposal

A   **B**

6) Again we can observe that it is possible to perform two types of disconnections on the nucleophilic center, thus obtaining two different types of products in each of the two proposals. Now the next step, our knowledge of chemistry, comes in, since in the first proposal the nitrogen of substance **A** (2,3-dimethylaniline) has a better nucleophilic character, since besides the methyl groups are electron donor groups, they are able to promote a certain degree of distortion of the planarity of the benzene ring, thus decreasing the participation of the nitrogen pair of electrons with the resonance of the ring, thus increasing its nucleophilicity. In substance **B** of the first proposal, (2-chlorobenzoic acid) we can observe that the leaving group is close to an electron-withdrawing group, which facilitates their exit, so this proposal is the most coherent and that used in the preparation of this drug.

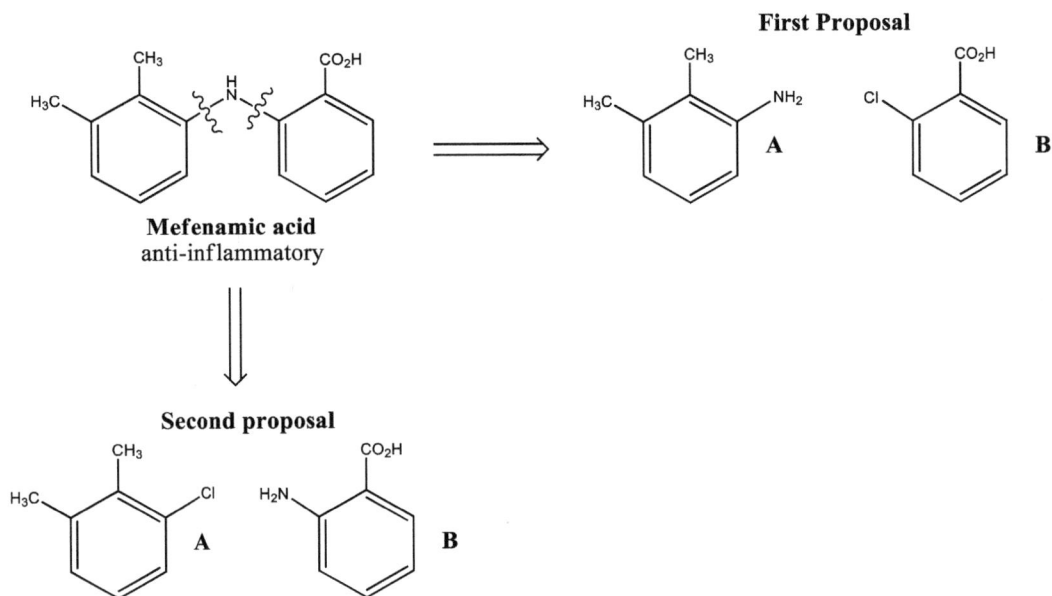

Mefenamic acid
anti-inflammatory

First Proposal

Second proposal

7)

Two reagents that can be simplified
retrosynthetic analysis.

Butan-2-ol which may have
been from 2-butanone.

ethinyl

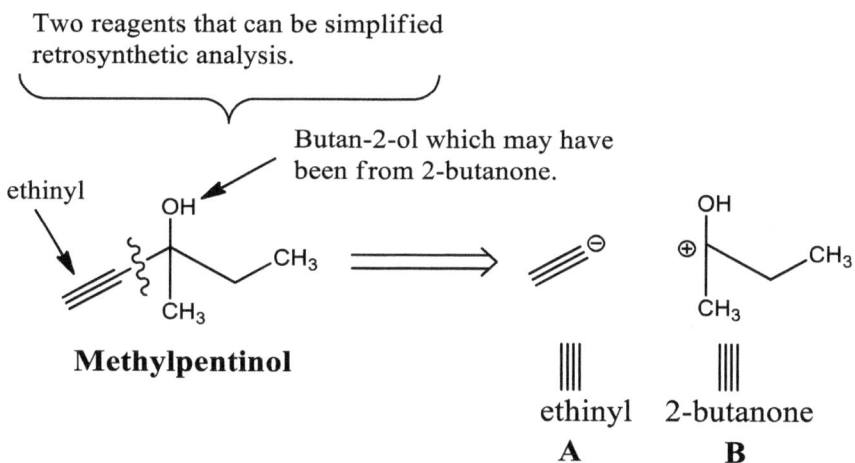

Methylpentinol

ethinyl   2-butanone

A   **B**

8) In this exercise, the recognition of substances and reagents again can significantly help in the preparation of this drug. However, if this is not possible, only the disconnections of the nucleophilic and electrophilic centers help us to propose the synthesis. We note that after the retrosynthetic analysis, it is clear that we need an amine and a dicarbonyl substance which may be a diester or a cyclic anhydride (substituted succinic anhydride).

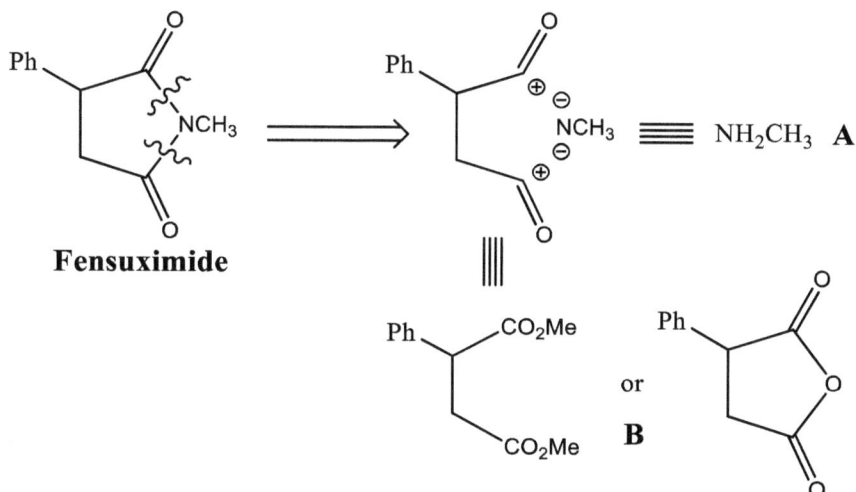

Fensuximide

NH_2CH_3 **A**

Ph CO_2Me

CO_2Me **B**

or

9) In a first analysis, we can observe that we have two nucleophilic centers (O and N) attached to an electrophilic center (CO), so this should be the starting point of our analysis. When making the disconnections, it is clear that we need a substance containing a carbon atom to join these two nucleophilic centers, phosgene.

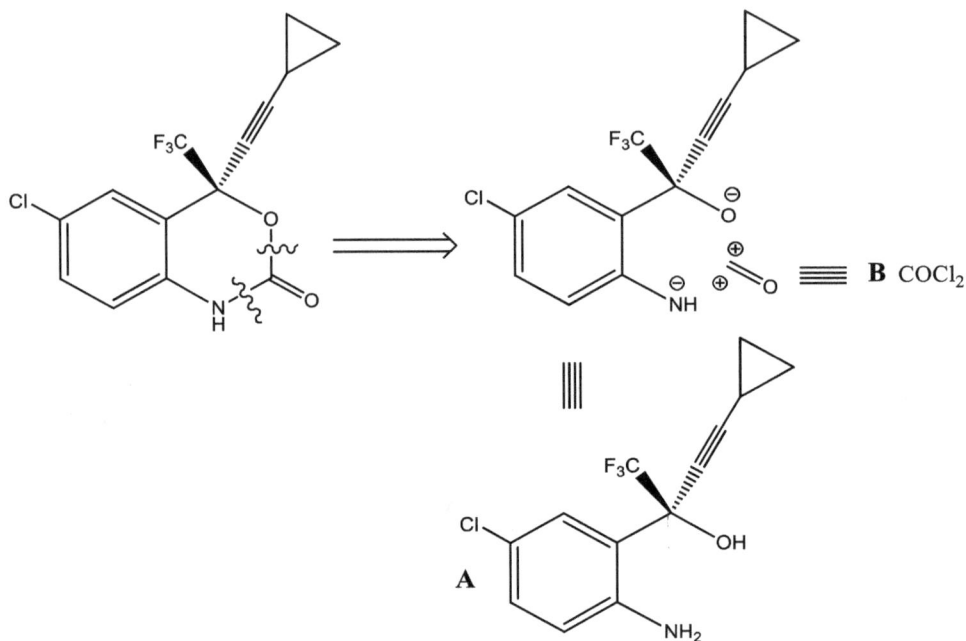

10) This exercise can easily mislead us; however, based on our chemical knowledge, we can avoid it. At first, as in exercise number **2**, we could propose the use of urea that would be attached to one of the aromatic rings. This urea reacts with a symmetrical aromatic ring (first proposal). However, this analysis, although correct, would be doubtful to occur, since the urea nitrogen is a non-nucleophilic base. This analysis shows us once again how important it is to know concepts, reactions, and chemical functions, as well as reagents and substances when evaluating and choosing a synthetic route. One well-known way of obtaining ureas and thioureas is to use their respective isocyanates and thioisocyanates. Always keep these transformations in mind when you come across these classes of substances. Indeed, this is not an apparent disconnection, especially if the reader does not know this kind of transformation. However, using our chemical expertise, we can see that the first proposal is not the best option.

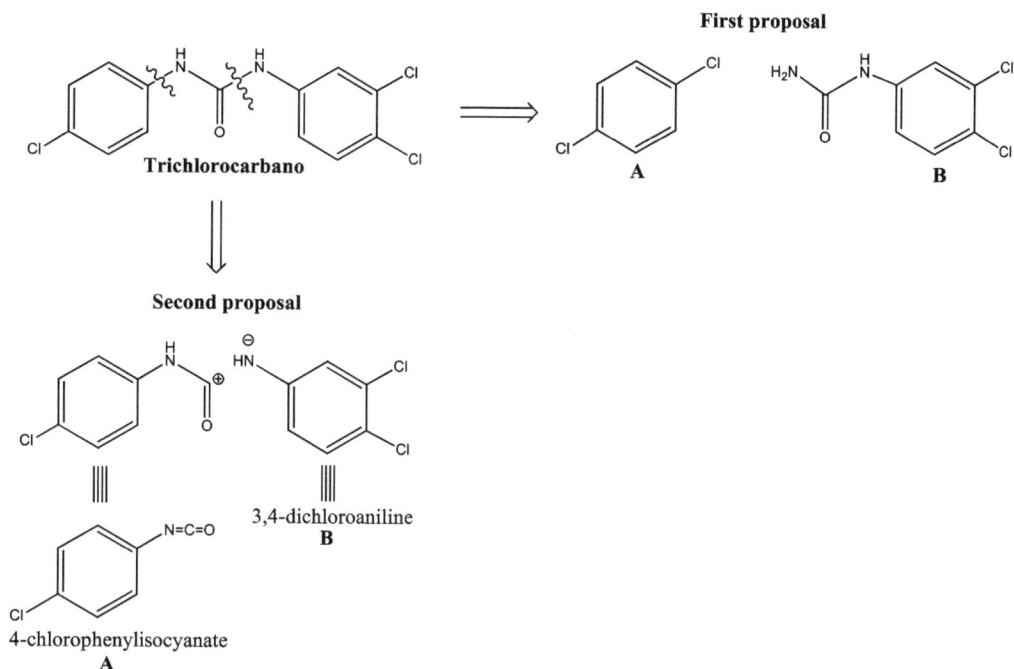

First proposal

Trichlorocarbano

Second proposal

3,4-dichloroaniline
B

4-chlorophenylisocyanate
A

EXERCISE 2 – EPOXIDES

Epoxides are very versatile substances and therefore very used in organic synthesis, being able to react with other elements with an economy of atoms, a principle widely used in green chemistry. Due to the importance and versatility of the epoxides in organic synthesis, the following exercises present drugs that can be synthesized using this class of substances; it is up to the reader to identify them, as well as the other synthetic equivalents (reagents and substances) that would participate in these synthesis.

Example

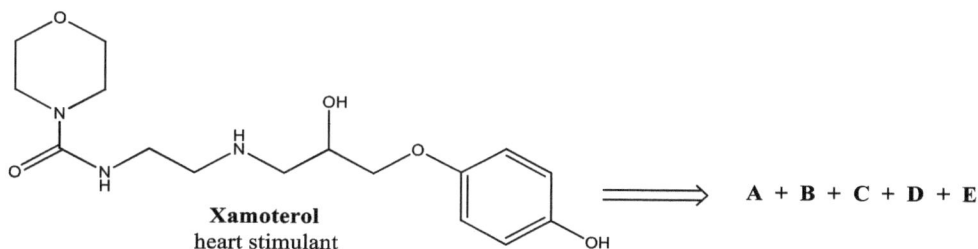

Xamoterol
heart stimulant

$A + B + C + D + E$

Answer

In the synthesis of the drug xamoterol, the disconnections, first of the nucleophilic centers, followed by the electrophilic centers, give us the synthons. This rational retrosynthetic analysis is a piece of valuable information for a consistent synthesis. Through these synthons, our work becomes much more comfortable to propose the synthesis and reagents used. It is noteworthy that, for the identification of the epoxide as a reagent, it is sufficient to keep in mind the presence of two or three carbon atoms neighboring an oxygen atom. It is also important to mention that there are other functional groups which have these characteristics, for example, ketones and alcohols. However, with only a quick analysis, the reader will identify the difference, since the epoxides will react with nucleophiles. For example, in the case of the target molecule in question, the reagent lies between two nucleophilic centers (oxygen and nitrogen) indicate the use of epichlorohydrin.

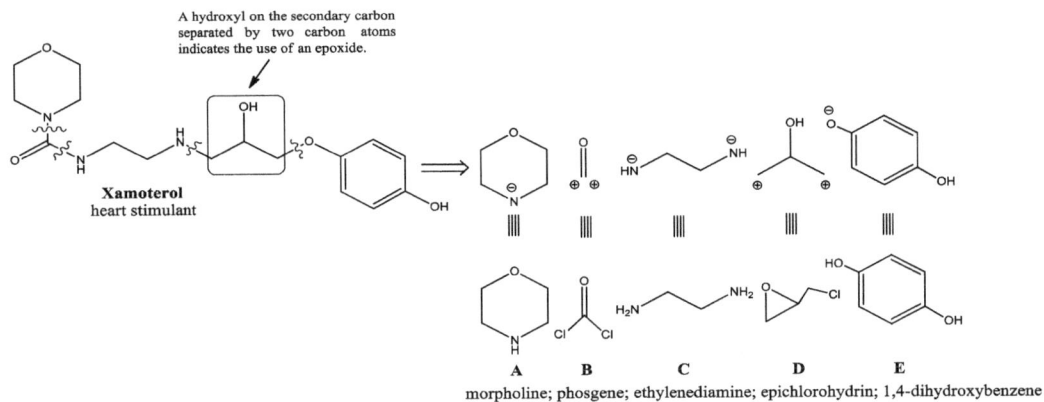

A hydroxyl on the secondary carbon separated by two carbon atoms indicates the use of an epoxide.

Xamoterol
heart stimulant

morpholine; phosgene; ethylenediamine; epichlorohydrin; 1,4-dihydroxybenzene

EXERCISES

1)

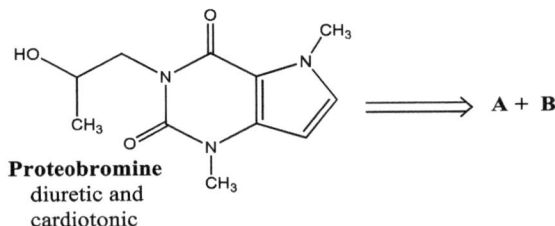

Proteobromine
diuretic and
cardiotonic

$\Longrightarrow$ **A + B**

2)

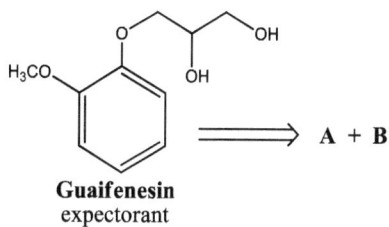

Guaifenesin
expectorant

$$\Longrightarrow \quad A + B$$

3)

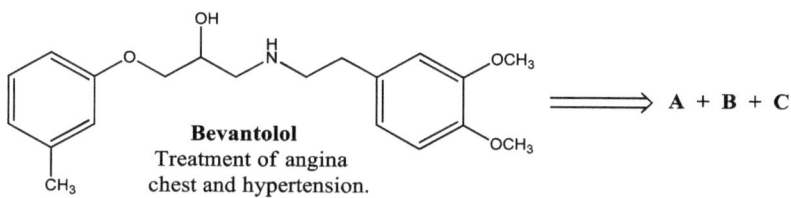

Bevantolol
Treatment of angina
chest and hypertension.

$$\Longrightarrow \quad A + B + C$$

4)

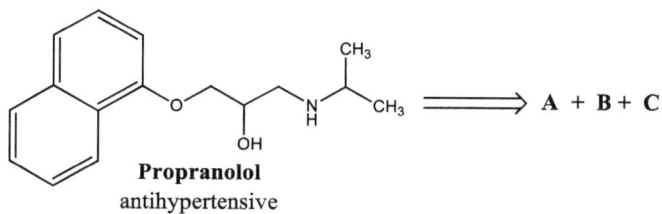

Propranolol
antihypertensive

$$\Longrightarrow \quad A + B + C$$

5)

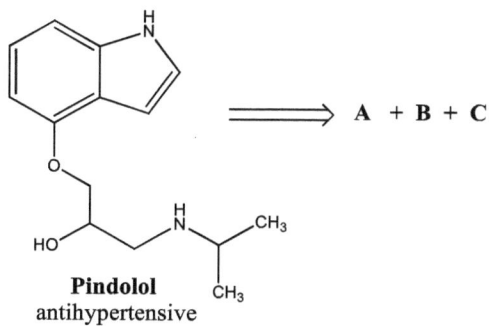

Pindolol
antihypertensive

$$\Longrightarrow \quad A + B + C$$

6)

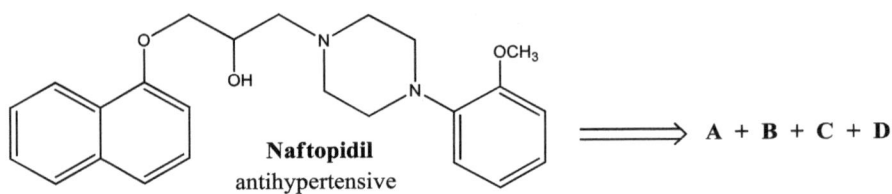

Naftopidil
antihypertensive

$\Longrightarrow$ **A + B + C + D**

7)

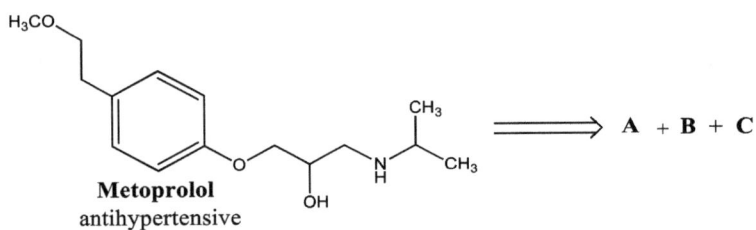

Metoprolol
antihypertensive

$\Longrightarrow$ **A + B + C**

8)

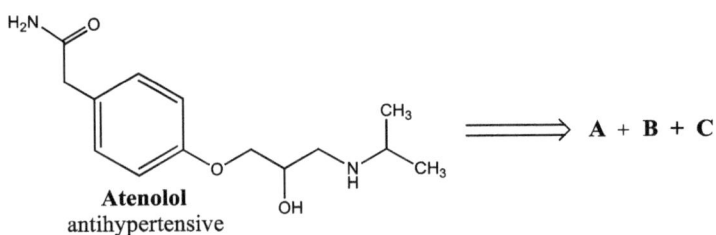

Atenolol
antihypertensive

$\Longrightarrow$ **A + B + C**

9)

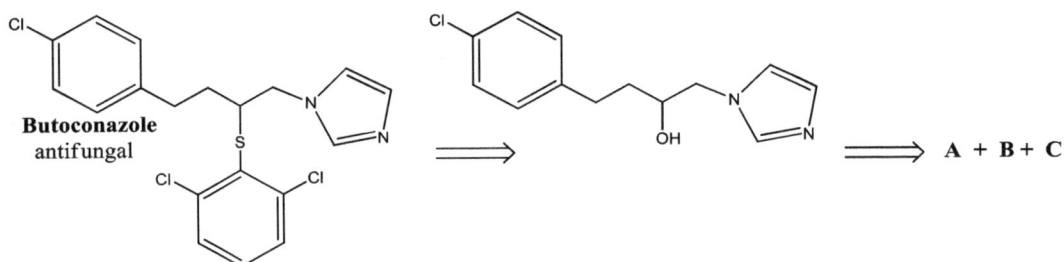

Butoconazole
antifungal

$\Longrightarrow$ $\Longrightarrow$ **A + B + C**

10)

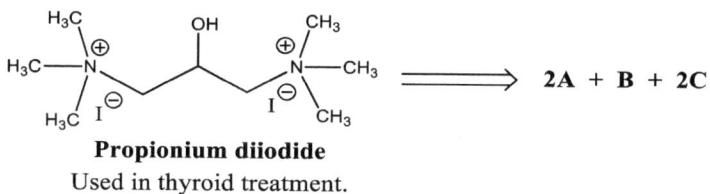

Propionium diiodide
Used in thyroid treatment.

$$2A + B + 2C$$

Answers

1)

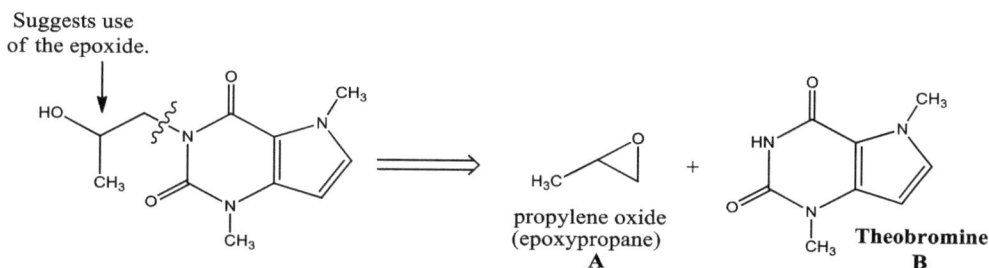

Suggests use of the epoxide.

propylene oxide
(epoxypropane)
A

Theobromine
B

2) A = glycidol; **B** = 2-methoxiphenol.

3)

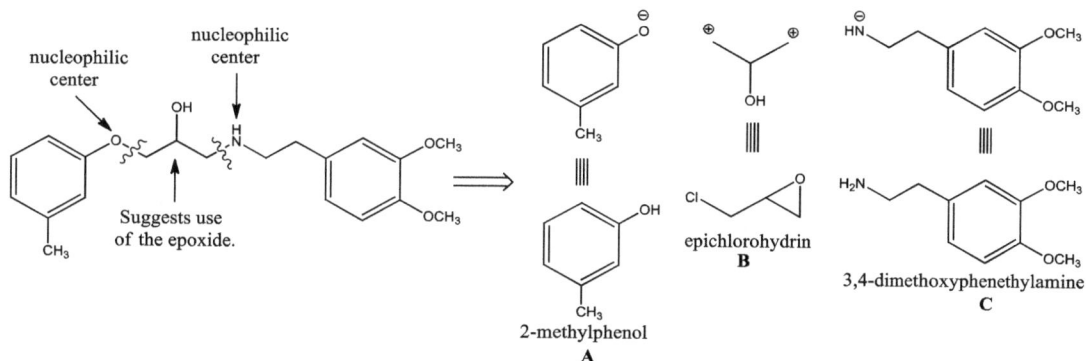

nucleophilic center

nucleophilic center

Suggests use of the epoxide.

2-methylphenol
A

epichlorohydrin
B

3,4-dimethoxyphenethylamine
C

4) A = β-naphthol; **B** = epichlorohydrin; **C** = isopropylamine.

5) A = 4-hidroxi-indol; **B** = epichlorohydrin; **C** = isopropylamine.

6) **A** = β-naftol; **B** = epichlorohydrin; **C** = piperazine; **D** = 1-fluoro-2-methoxybenzene.

7)

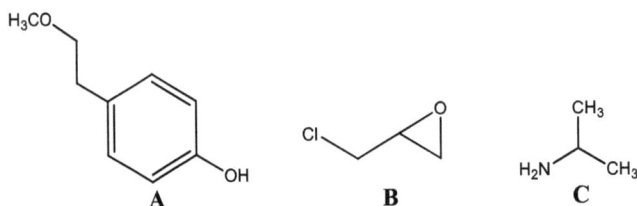

8)

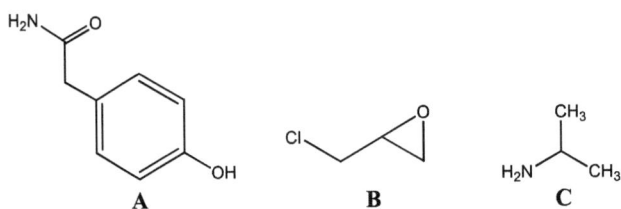

9)

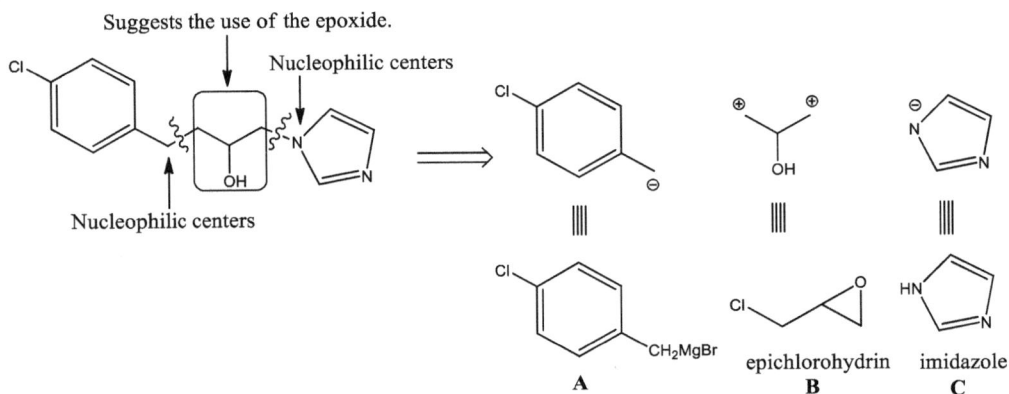

10) A = dimethylamine; **B** = epichlorohydrin; **C** = methyl iodide.

EXERCISE 3 – PIPERAZINE AND RELATED SUBSTANCES

Piperazine (Fig. **4**) is a saturated heterocyclic substance containing two nitrogen atoms in its structure at positions 1 and 4 and used as anthelmintics since 1953. Anthelmintics are also known as vermicides or vermifuges and are drugs capable of acting in a localized manner, eliminating worms from the gastrointestinal tract.

These compounds are also able to combat the helminths that invade other organs and tissues. In addition to this activity, piperazine and its related substances (Fig. **4**) have a wide variety of different activities, being widely used in the development of new drugs as well as being present in several drugs available in the market. The importance of piperazine in medicinal chemistry is due to several aspects, such as increased solubility, basicity, the formation of hydrogen bonding, conformation, and the ability to be intercalated between two substances. Due to the importance of piperazine and related substances, the following exercises present drugs that have these molecules in their structure. These exercises are to perform the retrosynthetic analysis of each of these drugs, identifying their respective synthons, substances, and reagents.

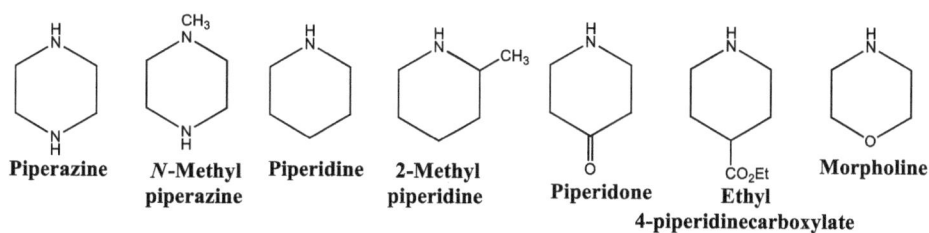

Fig. (4). Piperazine and related substances.

Examples

a)

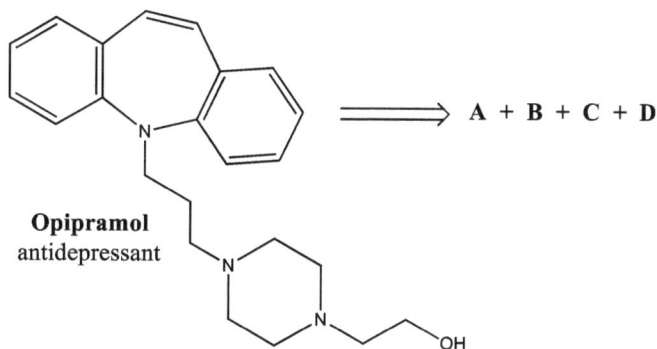

Opipramol
antidepressant

b)

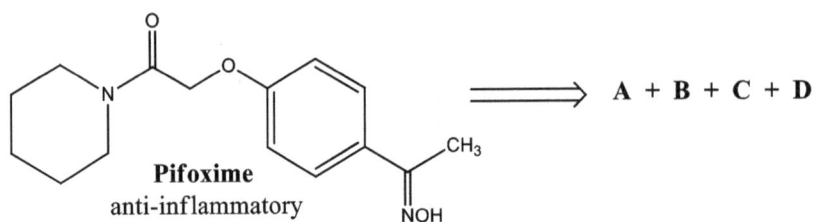

Pifoxime
anti-inflammatory

$\Longrightarrow$ **A + B + C + D**

Answers

a) The drug opipramol is an antidepressant widely used in Europe having a tricyclic (dibenzoazepine) system and an *N, N'*-disubstituted piperazine. In general, nitrogen atoms are nucleophilic centers, and it is important to think and try to make disconnections in these bonds. These nucleophilic centers will be related to synthons with a positive charge and will consequently facilitate to found the respective reagent.

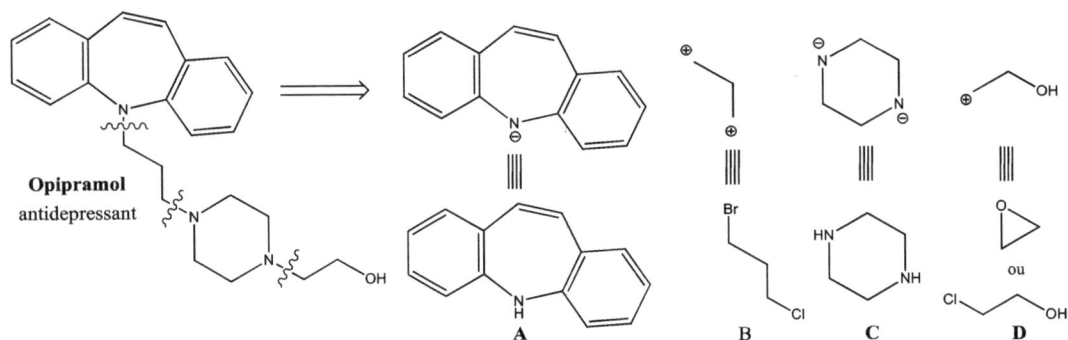

Opipramol
antidepressant

b) By the retrosynthetic analysis of the anti-inflammatory pifoxime, we have 4 synthons, which are synthetic equivalents of piperidine (**A**), chloroacetyl chloride (**B**), 4-hydroxyacetophenone (**C**) and hydroxylamine (**D**). Note that the identification of the substances used in the synthesis of the target molecule is easy when we rely on the disconnections since as nitrogen and oxygen are essentially nucleophilic centers, the neighboring atoms will consequently be electrophilic centers.

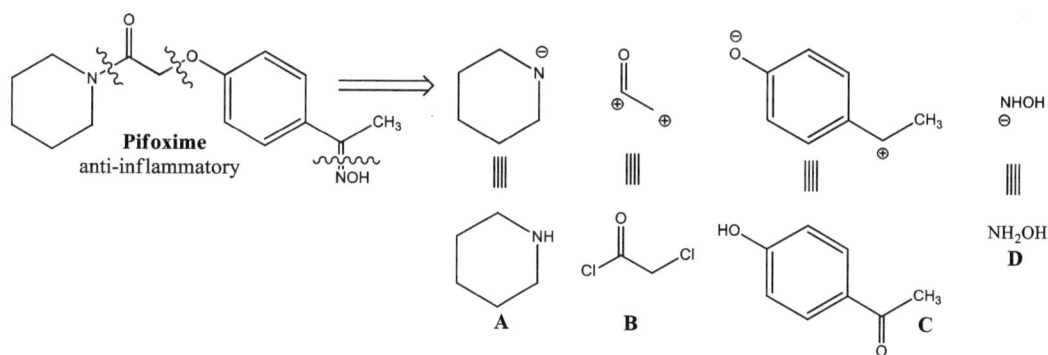

Pifoxime
anti-inflammatory

A   B   C   D

EXERCISES

1)

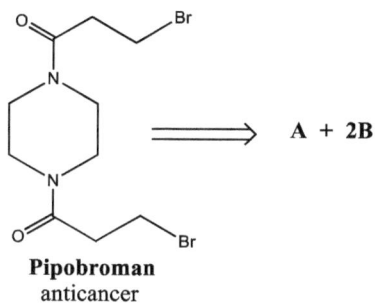

Pipobroman
anticancer

$$\Longrightarrow \quad A + 2B$$

2)

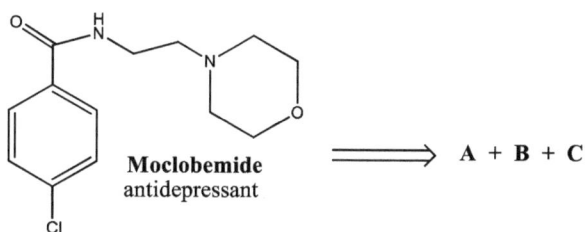

Moclobemide
antidepressant

$$\Longrightarrow \quad A + B + C$$

3)

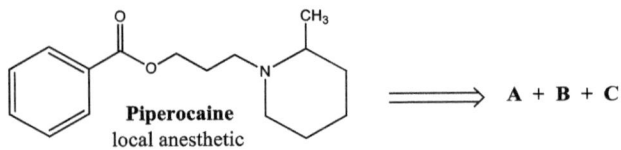

Piperocaine
local anesthetic

$\Longrightarrow$ **A + B + C**

4)

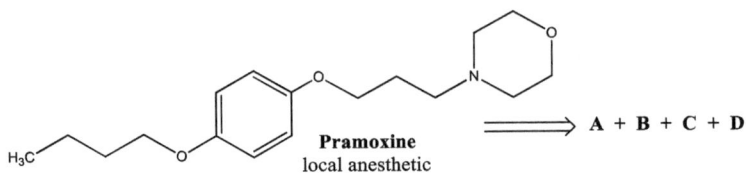

Pramoxine
local anesthetic

$\Longrightarrow$ **A + B + C + D**

5)

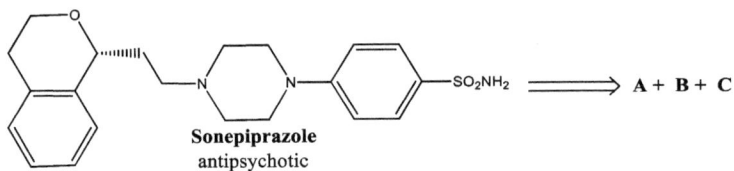

Sonepiprazole
antipsychotic

$\Longrightarrow$ **A + B + C**

6)

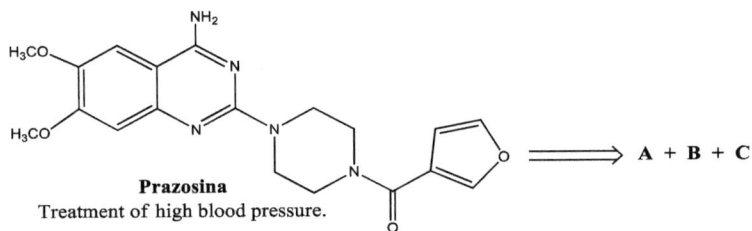

Prazosina
Treatment of high blood pressure.

$\Longrightarrow$ **A + B + C**

7)

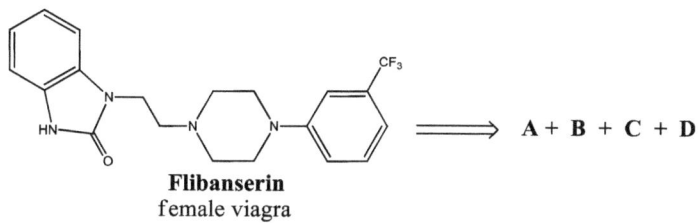

Flibanserin
female viagra

$$\Longrightarrow \mathbf{A + B + C + D}$$

8)

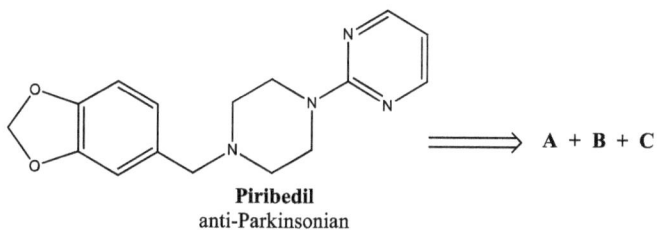

Piribedil
anti-Parkinsonian

$$\Longrightarrow \mathbf{A + B + C}$$

9)

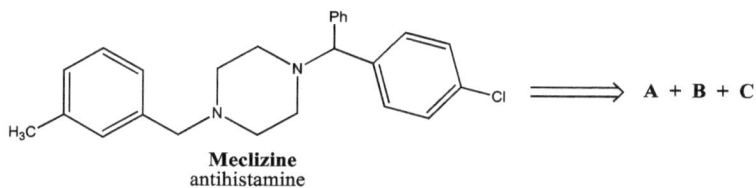

Meclizine
antihistamine

$$\Longrightarrow \mathbf{A + B + C}$$

10)

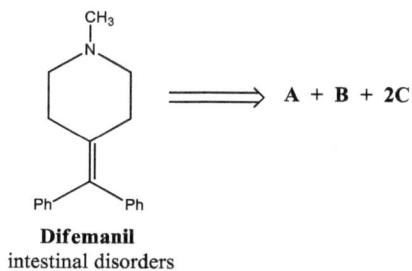

Difemanil
intestinal disorders

$$\Longrightarrow \mathbf{A + B + 2C}$$

11)

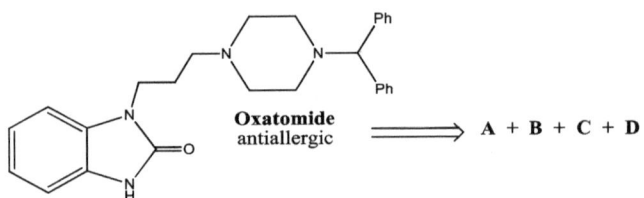

Oxatomide
antiallergic $\Longrightarrow$ **A + B + C + D**

12)

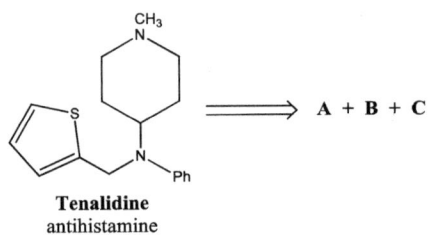

$\Longrightarrow$ **A + B + C**

Tenalidine
antihistamine

13)

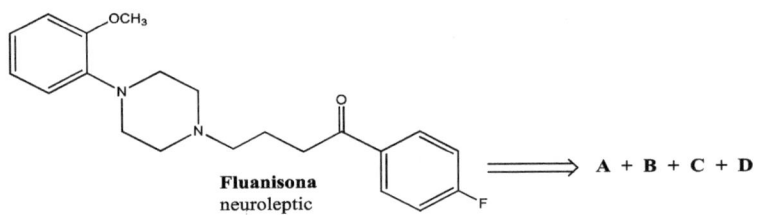

Fluanisona
neuroleptic $\Longrightarrow$ **A + B + C + D**

14)

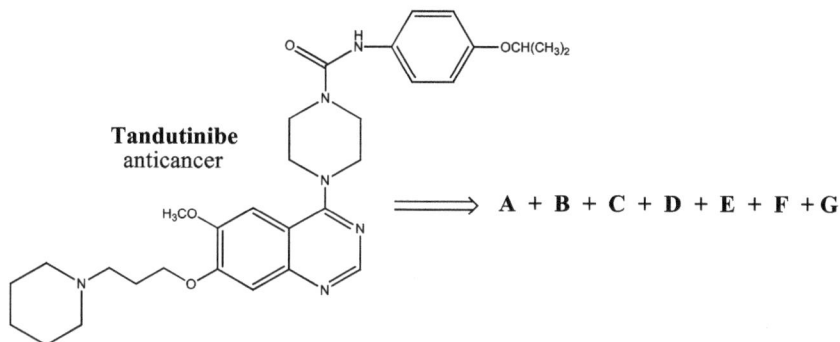

Tandutinibe
anticancer

$$\Longrightarrow \quad A + B + C + D + E + F + G$$

Answers

1) A = piperazine; **B** = 3-bromopropionyl chloride.

2)

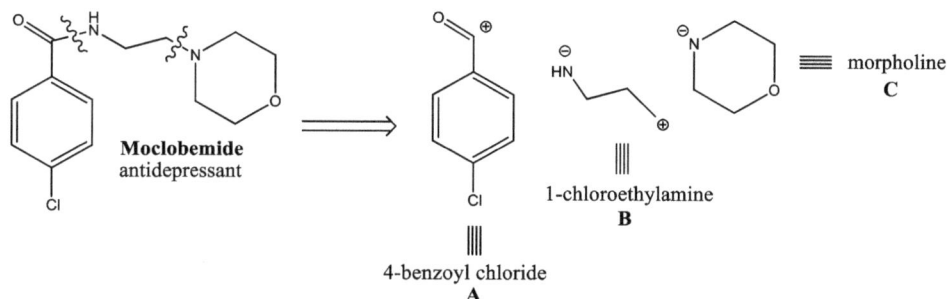

Moclobemide
antidepressant

morpholine
C

4-benzoyl chloride
A

1-chloroethylamine
B

3) A = benzoyl chloride; **B** = 3-chloropropanol; **C** = 2-methylpiperidine.

4) A = *n*-butyl bromide; **B** = 1,4-dihydroxybenzene (hydroquinone); **C** = 1-bromo-3-chloropropane; D = morpholine.

5)

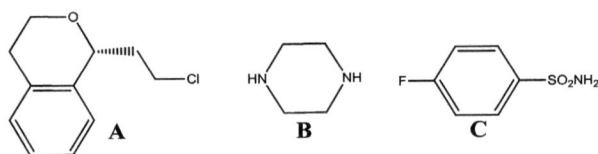

6)

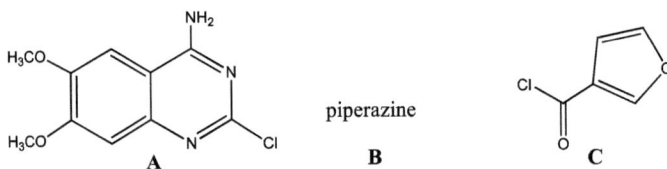

7)

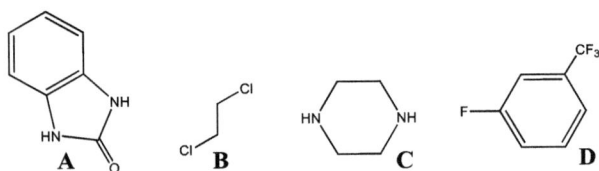

8)

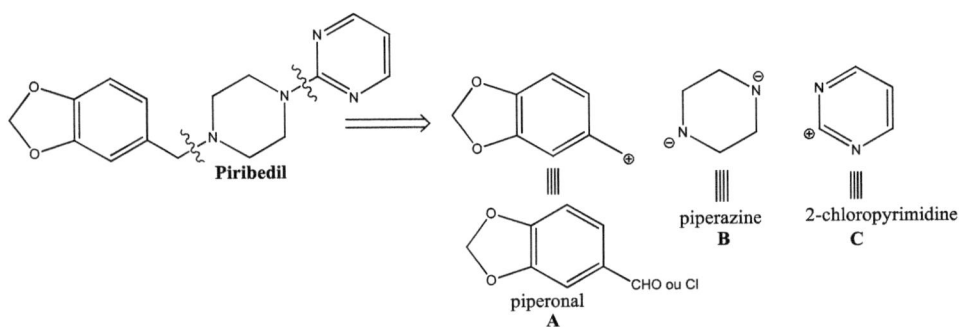

9) A = 3-methylbenzaldehyde; **B** = piperazine; **C** = *p*-chlorobenzhydryl chloride.

10) The critical step of the retrosynthetic analysis of this target molecule is the tetrasubstituted double bond. A transformation used, when we want to obtain double substituted bonds, is the nucleophilic addition to carbonyls followed by their dehydration. Thus, we have the disconnection in two different ways, the first using the ester function as the electrophilic center and the second employing the ketone (benzophenone). However, by analysis of the fragments and chemical knowledge; a correct choice is an ester group.

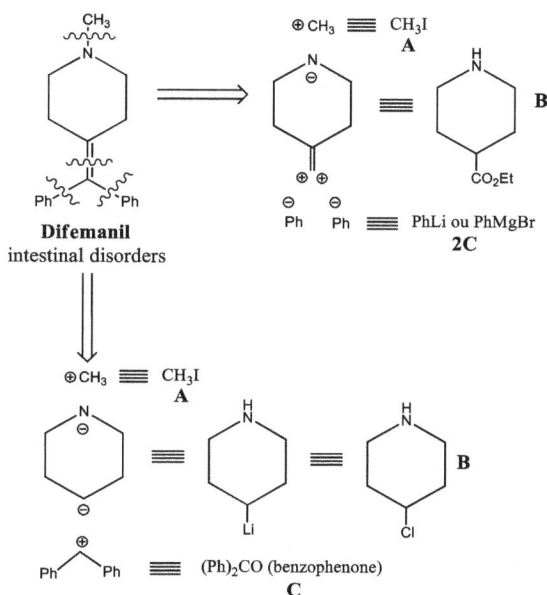

Difemanil
intestinal disorders

11) A = 2-benzimidazolone; **B** = 1-bromo-3-chloropropane; **C** = piperazine; **D** = benzhydryl chloride (synthesized from benzophenone).

12)

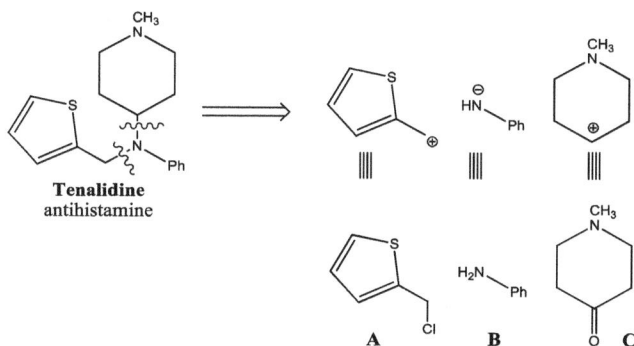

Tenalidine
antihistamine

13)

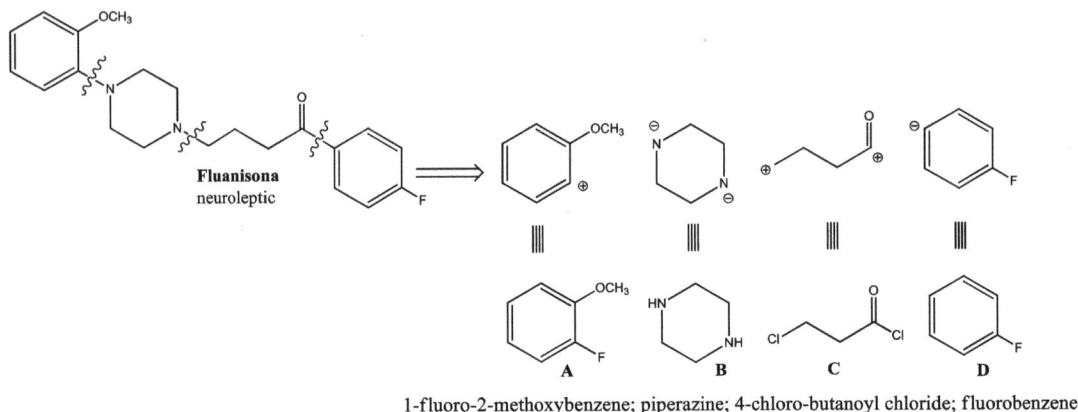

Fluanisona
neuroleptic

1-fluoro-2-methoxybenzene; piperazine; 4-chloro-butanoyl chloride; fluorobenzene

14) We can observe that, when we perform the disconnections of the target molecule, it is possible to identify the nucleophilic centers, located in the nitrogen and oxygen and, consequently, the electrophilic centers. Therefore, we have reasonable indications of the synthons, as well as the reagents and substances used.

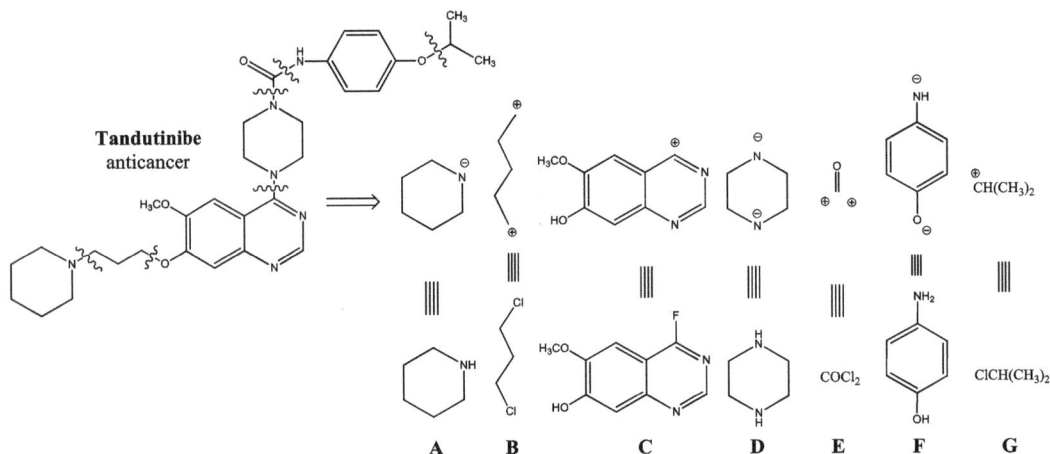

Tandutinibe
anticancer

EXERCISE 4 – ALDEHYDES

The aldehyde chemical function presents a great versatility in organic synthesis, since it is capable of being used in a large number of chemical transformations such as the production of alcohols, ethers, carboxylic acids, esters, oximes, alkenes, among others. The aldehydes can also be used in both carbon-carbon bond formation and in the preparation of various heteroaromatic and cyclic systems. Due to the importance of aldehydes function, the exercises below will use this class, reagents, and substances to synthesize drugs and bioactive compounds. Based on the concepts of retrosynthesis and synthon, it is up to the reader to identify aldehyde, reagents, and substances in the target molecules.

Examples

a)

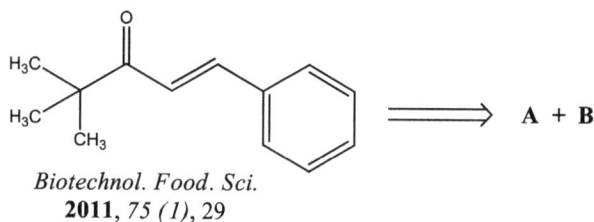

Biotechnol. Food. Sci.
2011, *75 (1)*, 29

$\Longrightarrow$ **A + B**

b)

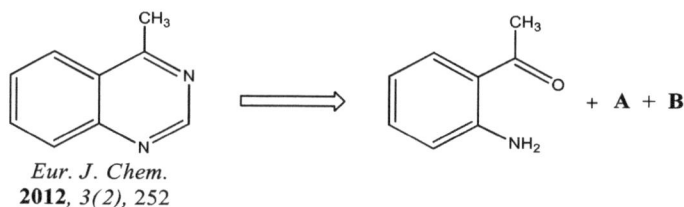

Eur. J. Chem.
2012, *3(2)*, 252

$\Longrightarrow$ + **A + B**

c)

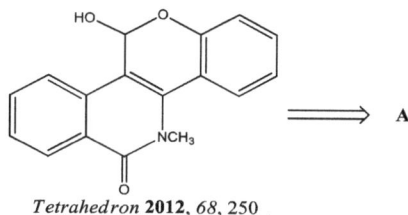

Tetrahedron **2012**, *68*, 250

$\Longrightarrow$ **A**

Answers

a) As we can see, in disconnection, we know that α-carbonyl carbons are synthons that have negative charges and ketone and aldehyde carbons are synthons that have positive charges. Thus, the substance formed, there must be an aldol condensation followed by dehydration of the aldol (Claisen-Schmidt reaction) between ketone **A** and benzaldehyde **B**.

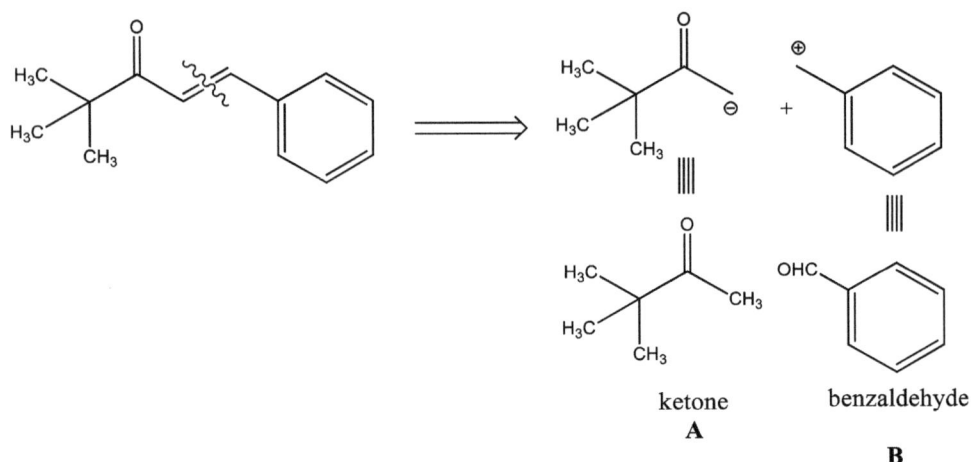

ketone
A

benzaldehyde

B

b) From the information provided, the synthesis of the quinazoline nucleus was obtained using 2-aminoacetophenone plus two substances. By the structure of the target molecule, we can observe that there was an introduction of a nitrogen atom and loss of an oxygen atom, besides the addition of one more carbon atom. Then, it is easy to conclude that the use of formaldehyde (**A**) and ammonia (**B**) are necessary components. However, another way to solve the problem is by retrosynthetic analysis.

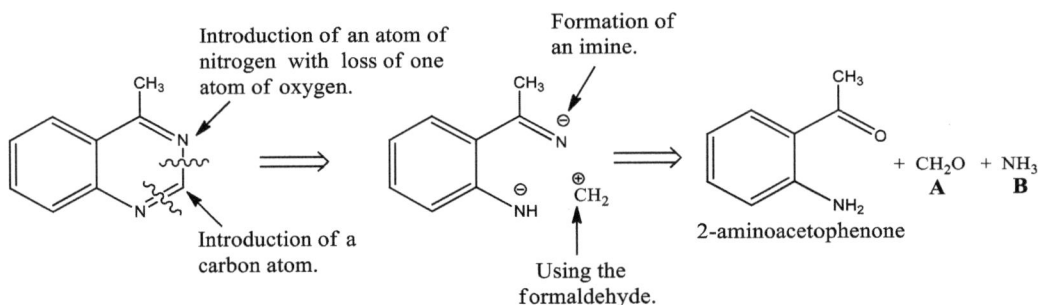

Introduction of an atom of nitrogen with loss of one atom of oxygen.

Formation of an imine.

Introduction of a carbon atom.

Using the formaldehyde.

2-aminoacetophenone

+ CH_2O + NH_3
A   **B**

c) In the analysis of the target molecule, we can quickly identify a nucleophilic center. With this, it is easy to identify the aldehyde function and consequently solve the exercise. Another way of solving the problem would be the observation that we have a chemical function known as hemiacetal. Hemiacetais, as well as hemiketals, are substances obtained by the reaction between a certain alcohol and a carbonylated substance, being hemiacetals (alcohol + aldehyde) and hemiketals (alcohol + ketone).

Hemiacetal

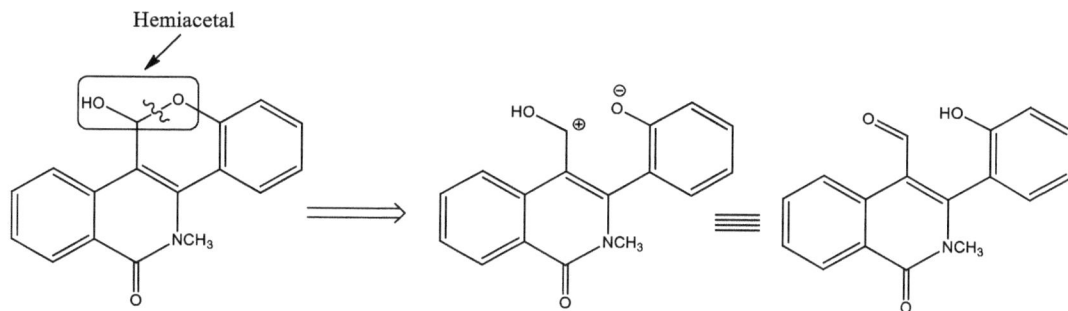

EXERCISES

1)

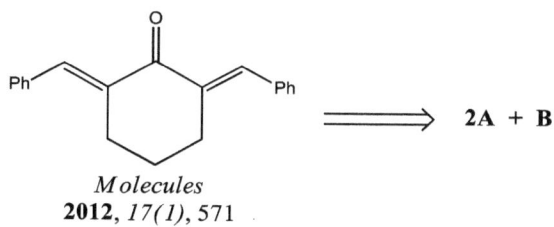

$$\Longrightarrow \quad \textbf{2A} + \textbf{B}$$

Molecules
2012, *17(1)*, 571

2)

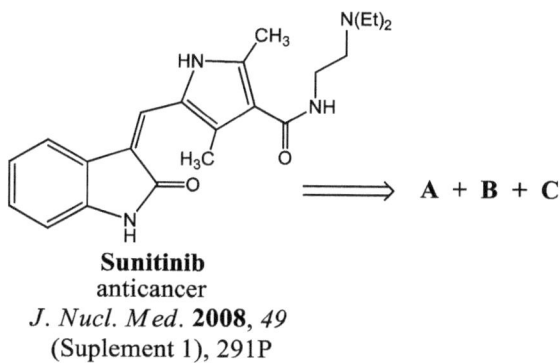

$$\Longrightarrow \quad \textbf{A} + \textbf{B} + \textbf{C}$$

Sunitinib
anticancer
J. Nucl. Med. **2008**, *49*
(Suplement 1), 291P

3)

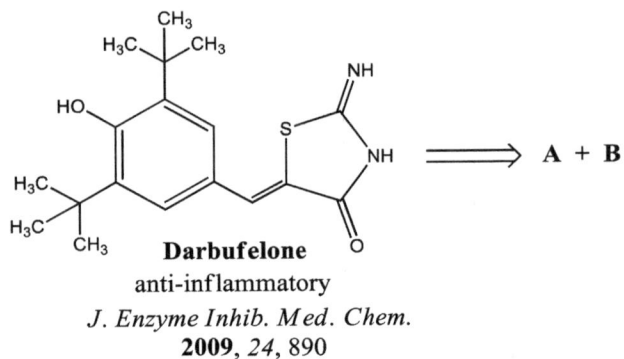

Darbufelone
anti-inflammatory
J. Enzyme Inhib. Med. Chem.
2009, *24*, 890

$\Longrightarrow$ **A + B**

4)

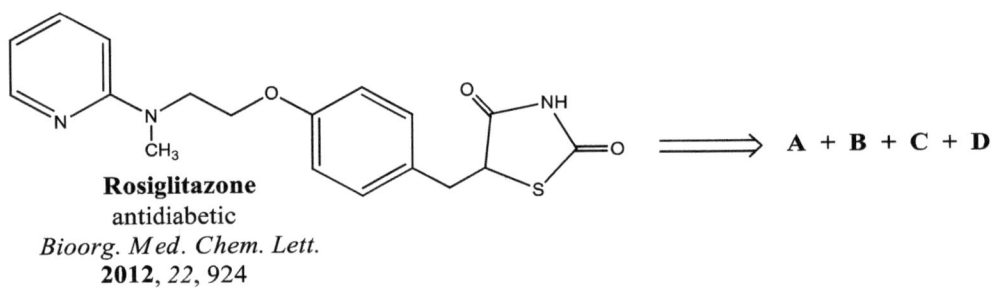

Rosiglitazone
antidiabetic
Bioorg. Med. Chem. Lett.
2012, *22*, 924

$\Longrightarrow$ **A + B + C + D**

5)

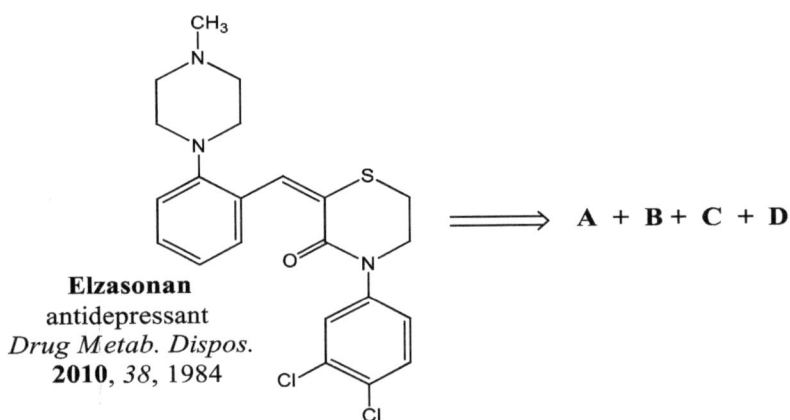

Elzasonan
antidepressant
Drug Metab. Dispos.
2010, *38*, 1984

$\Longrightarrow$ **A + B + C + D**

6)

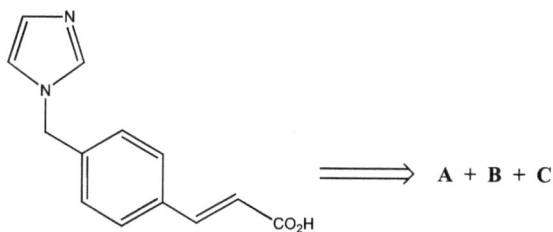

Ozagrel
antiplatelet agent
Chin. Chem. Lett.
2006, *17*, 180

$$\Longrightarrow \quad A + B + C$$

7)

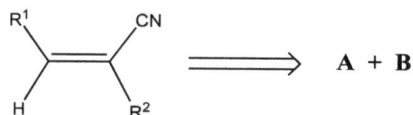

Synth. Commun.
2009, *39*, 2001

$$\Longrightarrow \quad A + B$$

8)

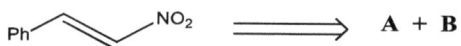

Tetrahedron Lett.
2012, *53*, 2980

$$\Longrightarrow \quad A + B$$

9)

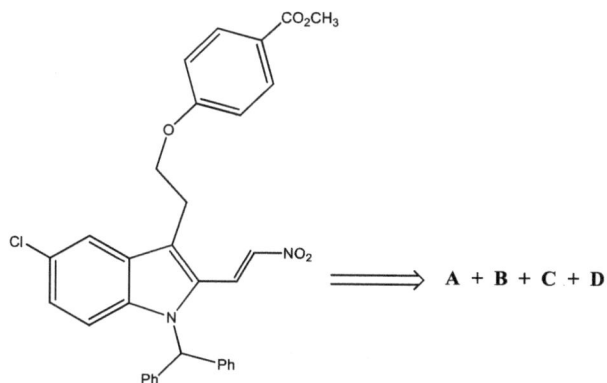

Key Intermediate of **Ecopladib**.
Asthma treatment.
Tetrahedron **2008**, *64*, 7871

$$\Longrightarrow \quad A + B + C + D$$

10)

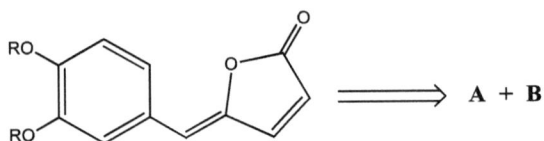

Synth. Commun. **2010**, *40*, 3346

11)

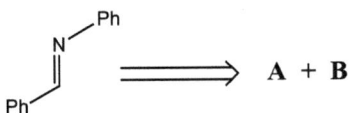

Green. Chem. Lett. Rev.
2011, *4(3)*, 195

12)

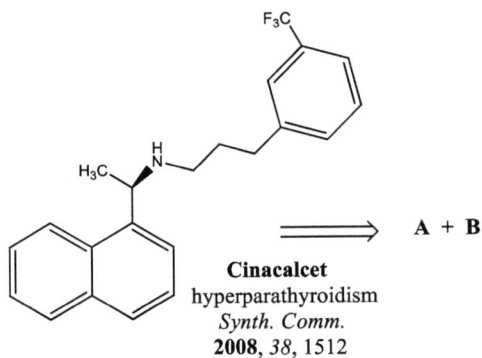

Cinacalcet
hyperparathyroidism
Synth. Comm.
2008, *38*, 1512

13)

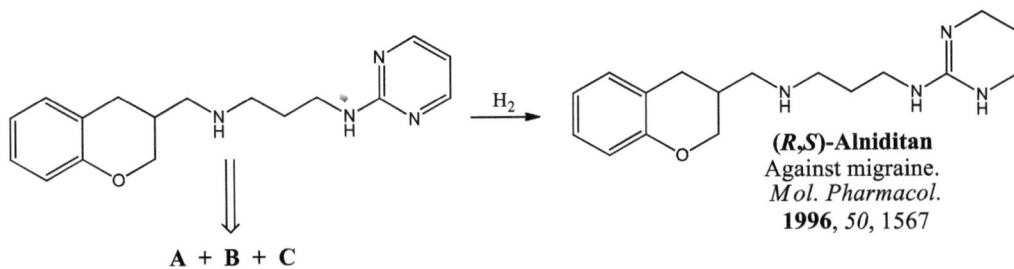

(R,S)-Alniditan
Against migraine.
Mol. Pharmacol.
1996, *50*, 1567

A + B + C

14)

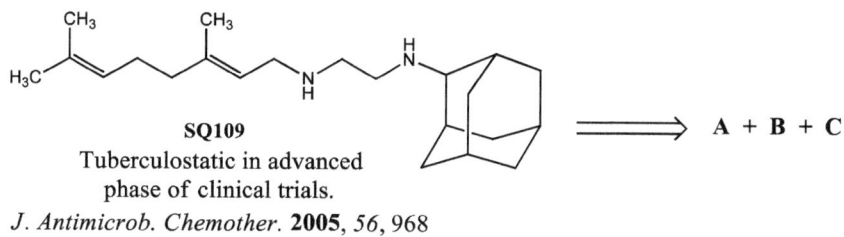

SQ109
Tuberculostatic in advanced
phase of clinical trials.
J. Antimicrob. Chemother. **2005**, *56*, 968

$\Longrightarrow$ **A + B + C**

15)

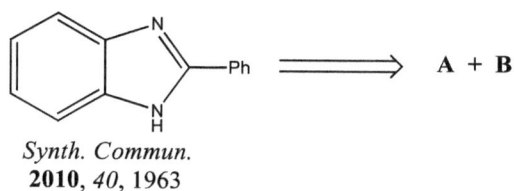

Synth. Commun.
2010, *40*, 1963

$\Longrightarrow$ **A + B**

16)

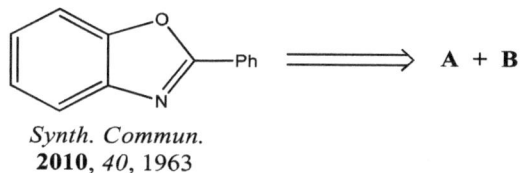

Synth. Commun.
2010, *40*, 1963

$\Longrightarrow$ **A + B**

17)

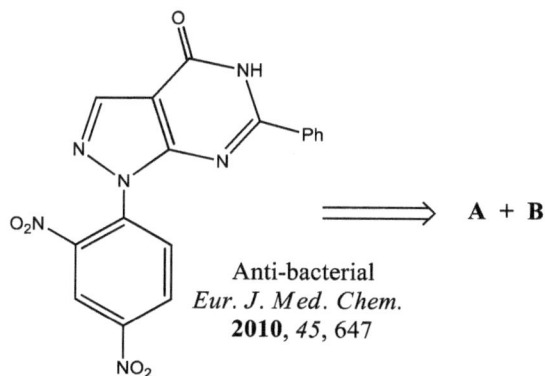

Anti-bacterial
Eur. J. Med. Chem.
2010, *45*, 647

$\Longrightarrow$ **A + B**

18)

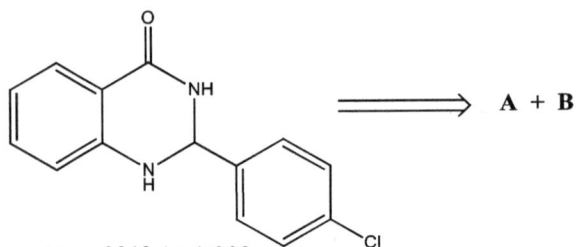

Arkivoc **2012** (*viii*) 308

$\Longrightarrow$ **A + B**

19)

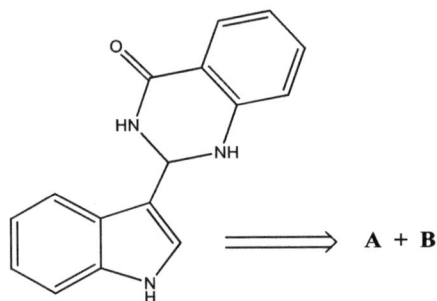

Tetrahedron Lett.
2012, *53*, 2222

$\Longrightarrow$ **A + B**

20)

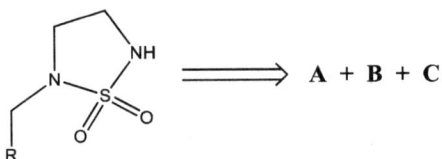

Viral gastroenteritis
Bioorg. Med. Chem.
2011, *19*, 5975

$\Longrightarrow$ **A + B + C**

21)

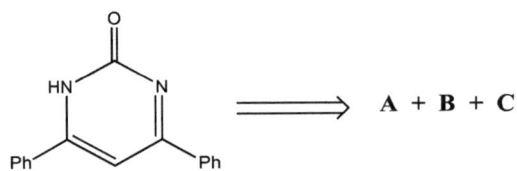

Synth. Commun.
2010, *40*, 1256

$\Longrightarrow$ **A + B + C**

22)

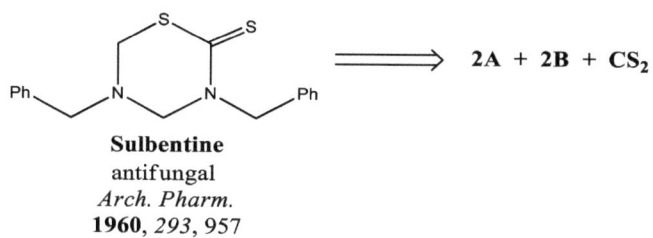

Sulbentine
antifungal
Arch. Pharm.
1960, *293*, 957

$$\Longrightarrow \quad 2A \;+\; 2B \;+\; CS_2$$

Answers

1) A = cyclohexanone; **B** = benzaldehyde.

2)

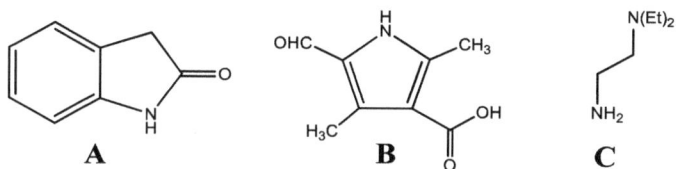

3)

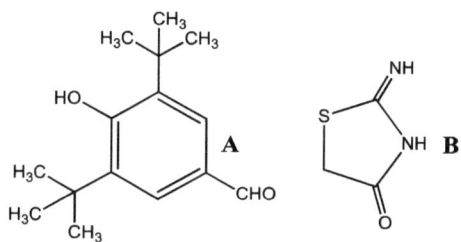

4)

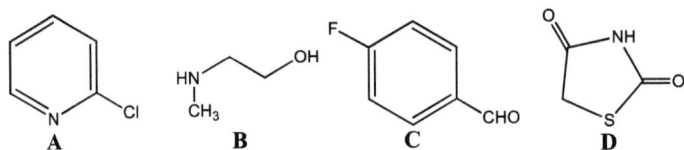

or

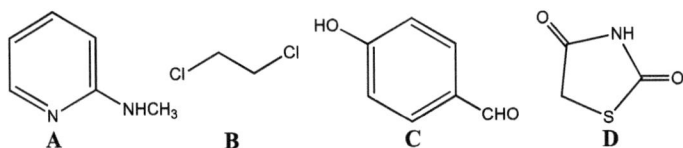

5)

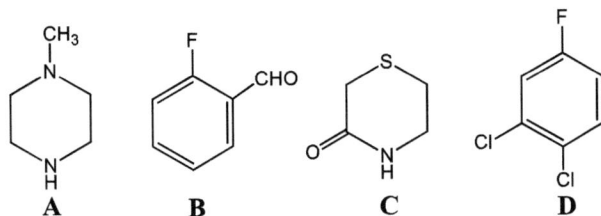

6)

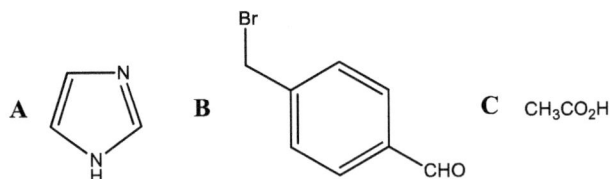

7) A = R^{2}CH$_2$CN; **B** = R^1-CHO.

8) A = benzaldehyde; **B** = nitromethane.

9)

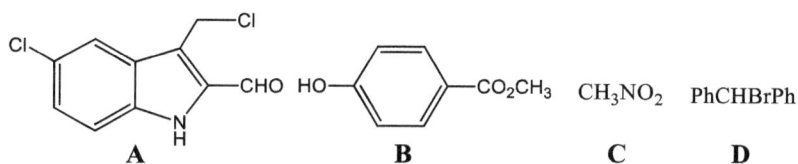

A   B   C   D

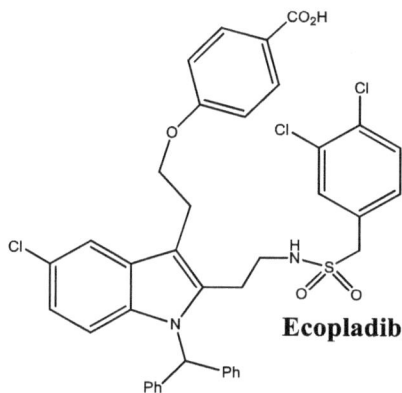

Ecopladib

10)

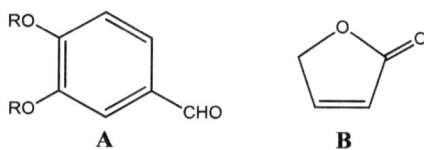

A   B

11) A = benzaldehyde; **B** = aniline.

12)

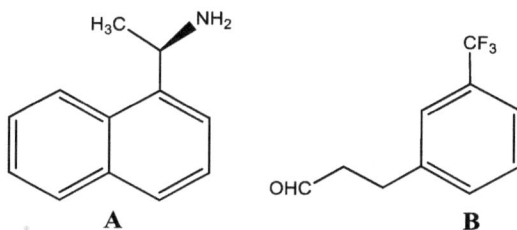

A   B

13)

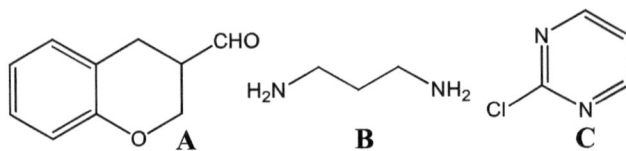

14)

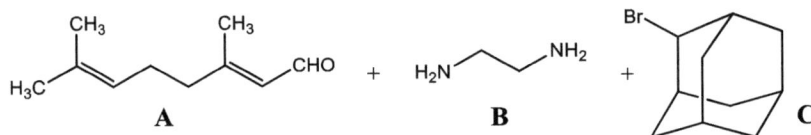

15) A = *ortho*-phenylenediamine; **B** = benzaldehyde.

16) A = 2-aminophenol; **B** = benzaldehyde.

17)

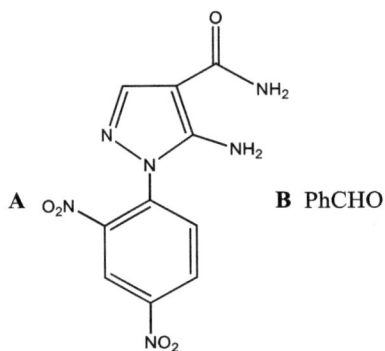

B PhCHO

18)

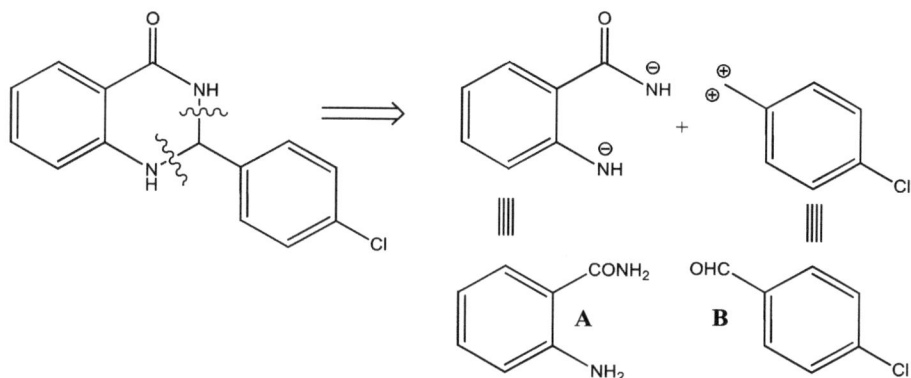

19)

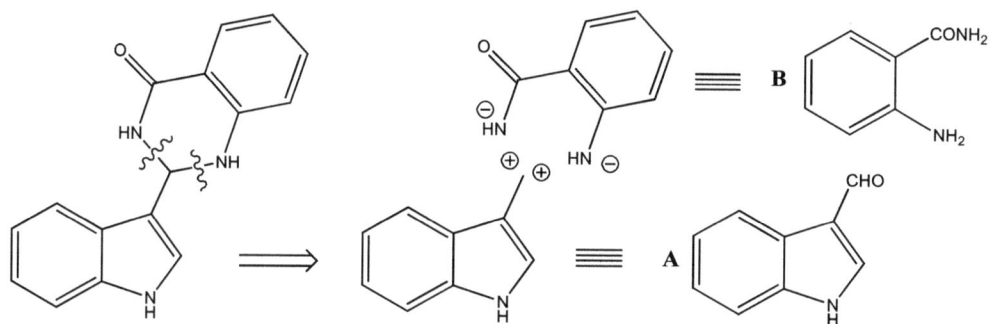

20)

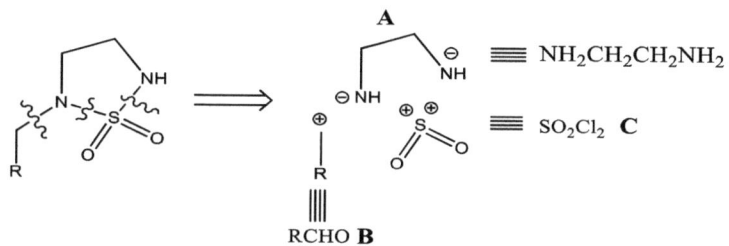

21)

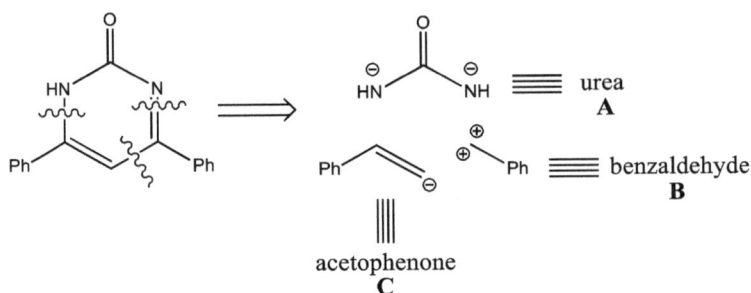

urea
A

benzaldehyde
B

acetophenone
C

22) Through the disconnections of the drug, we have two important information. First, the synthon coming from the carbon disulfide and, the second both components **A** and **B** are in the ratio of two moles. From the nucleophilic centers, and our knowledge of reagents, we concluded that one of the substances in question is benzylamine, and the aldehyde in question is formaldehyde.

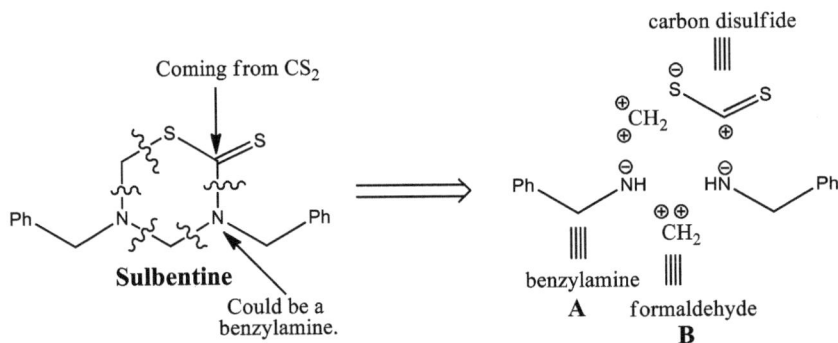

carbon disulfide

Coming from CS_2

Sulbentine
Could be a
benzylamine.

benzylamine
A   formaldehyde
B

CONCLUSION OF THE EXERCISES 1 TO 4 (RETROSYNTHETIC ANALYSIS)

The development of a new drug begins with the search, planning, design, and evaluation being necessary for its preparation in the laboratory. This initial step is a long and complicated process for a biologically active compound to become a drug. For this to happen, organic synthesis is of fundamental importance since, through it, we can obtain the most varied types of simple or complex substances for biological evaluation.

Concepts involving retrosynthesis were and are essential to the development of organic synthesis as science and art. Because of its crucial importance for the learning of organic synthesis, the first four exercises were based on this concept. It is worth mentioning that whenever possible, it is highly recommended that the reader employs these concepts.

REFERENCES

[1]   Kocienski, P. *Protecting Groups,* 4th ed.; Thieme, **2005**.

[2]   Wuts, P.G.M.; Greene, T.W. *Greene's Protective Groups in Organic Synthesis,* 4th ed.; John Wiley & Sons, Inc.: New Jersey, **2006**.

CHAPTER 2

Organizing Synthetic Routes and Identifying Reagents

Abstract: In this chapter, the exercises below are based on two themes. First, exercises based on the synthesis of several drugs and bioactive compounds used to treat various diseases. Therefore, several reaction conditions are presented. However, these conditions are disordered, and it is up to the reader to order them. The other theme is to identify reagents used for the preparation of a specific target molecule. This type of training is essential because it allows the reader to contact different types of reagents and chemical transformations and use them in a sequentially correct way.

Keywords: Bioactive compounds, Chemical transformations, Drugs, Exercises, Medicinal chemistry, Molecules, Reaction conditions, Reagents, Synthesis.

EXERCISES 5 TO 8

INTRODUCTION

After the exercises on retrosynthetic analysis, the reader is now invited to practice other essential aspects of a particular synthetic route. It is worth mentioning that the retrosynthetic analysis has to be allied to solid chemical knowledge. In this context, the identification of reactions, reagents, intermediates, and chemical transformations are essential. Considering that, different exercises in this book are intended to make the reader familiar with these themes.

Exercises **5** to **8** present some of these objectives, Exercise **5**, for example, proposes different syntheses with the reagents arranged incorrectly, and it is up to the reader to order them. Exercise **6** aims at identifying the reagents used and their correct order of use in the preparation of a given drug (target molecule). Exercise **7** has the objective of identifying the structure of the drugs, having information of the starting materials and reagents used. Exercise **8** presents a series of gaps in the synthesis of different drugs, and it is up to the reader to complete them. It is worth mentioning that it is highly recommended that the reader continues using the concept of retrosynthesis in the resolution of the exercises.

EXERCISE 5 - ORGANIZING SYNTHETIC ROUTES

In the exercises below, different synthetic bioactive substances are presented, including some drugs used to treat various diseases. However, the reactional conditions for obtaining them are disordered, and it is up to the reader to order them.

Example

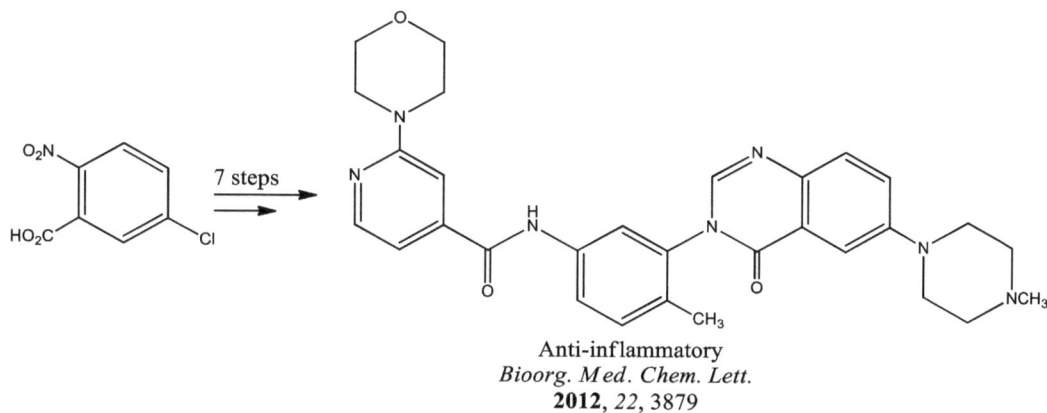

Anti-inflammatory
Bioorg. Med. Chem. Lett.
2012, *22*, 3879

(**a**) 4-chloro-2-chloropyridine acid chloride, (**b**) (OEt)$_3$CH, AcOH, EtOH, 70°C, 98%; NEt$_3$, CH$_2$Cl$_2$, rt, 91%; (**c**) morpholine, 100°C, 57%; (**d**) (COCl)$_2$, DMF, CH$_2$Cl$_2$, 0°C to rt; (**e**) 2-methyl-5-nitroaniline, NEt$_3$, CH$_2$Cl$_2$, rt (99% 2 steps); (**f**) H$_2$, 10% Pd/C, MeOH, rt, 80%; (**g**) *N*-methylpiperazine, 100°C, 83%.

Answer

However, it is worth emphasizing the importance of first doing a retrosynthetic analysis, because it will undoubtedly provide relevant information for solving the problem. In this context, the first question to ask is where the starting material in the target molecule is? It looks like that the fused nuclei **C** and **B** are derived from this raw material. The next step is to identify which transformations occur. We can highlight the following: a) conversion of the carboxylic acid into amide *via* formation of an acid chloride; b) coupling with the core **D**; c) synthesis of the nucleus **C** using a carbon atom and d) introduction of the *N*-methylpiperazine (nucleus **A**). The disconnections and coupling in the **D-F** rings are easy to see in conjunction with the reaction conditions shown. Thus, the ordered reaction sequence is as described below.

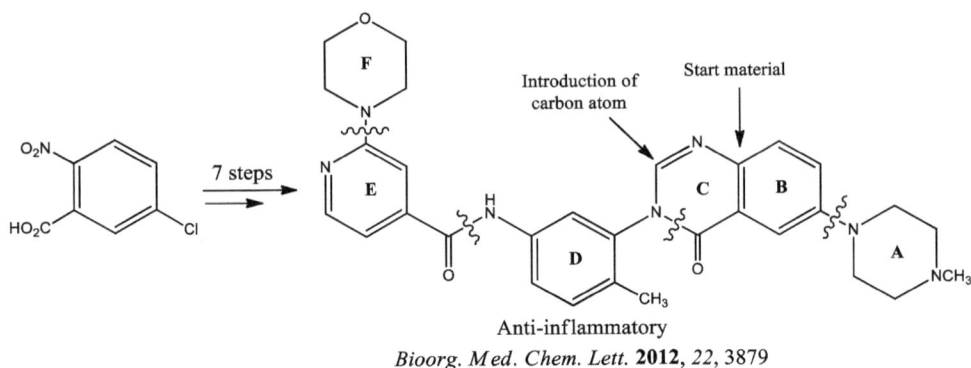

Anti-inflammatory
Bioorg. Med. Chem. Lett. **2012**, *22*, 3879

(**a**) (COCl)$_2$, DMF, CH$_2$Cl$_2$, 0°C at rt; (**b**) 2-methyl-5-nitroaniline, NEt$_3$, CH$_2$Cl$_2$, rt (99%; 2 steps); (**c**) *N*-methylpiperazine, 100°C, 83%; (**d**) H$_2$, 10% Pd/C, MeOH, rt, 80%; (**e**) (OEt)$_3$CH, AcOH, EtOH, 70°C, 98%; (**f**) 2-chloropyridine-4-carbonyl chloride, NEt$_3$, CH$_2$Cl$_2$, rt, 91%; (**g**) morpholine, 100°C, 57%.

EXERCISES

1)

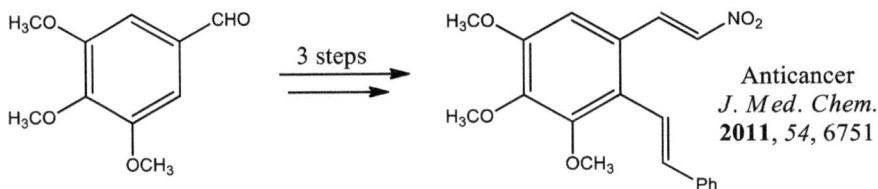

Anticancer
J. Med. Chem.
2011, *54*, 6751

(**a**) nitromethane and base (NH$_4$OAc), 100°C, 3h.; (**b**) Heck reaction with styrene; (**c**) Br$_2$, AcOH, rt.

2)

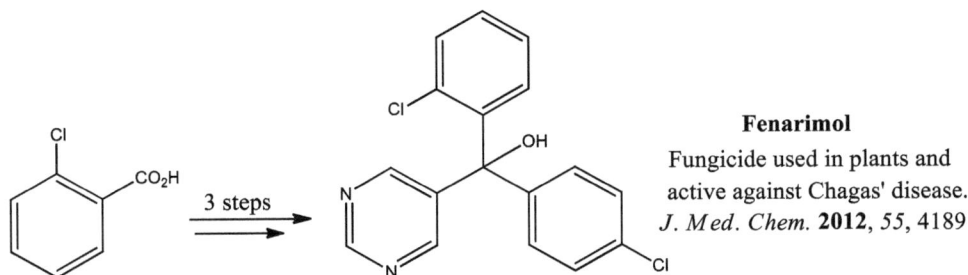

Fenarimol
Fungicide used in plants and active against Chagas' disease.
J. Med. Chem. **2012**, *55*, 4189

(**a**) 5-bromopyrimidine, *n*-BuLi, THF, -78°C; (**b**) thionyl chloride; (**c**) AlCl$_3$, PhCl.

3)

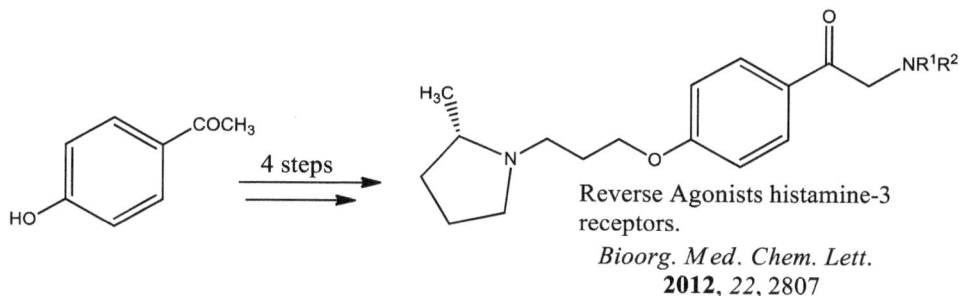

Reverse Agonists histamine-3 receptors.
Bioorg. Med. Chem. Lett.
2012, *22*, 2807

(**a**) R₂NH, K₂CO₃, MeCN, 80°C, >90%; (**b**) (*R*)-2-methylpyrrolidine. HCl, NaI, 2-butanone, reflux, 30%; (**c**) Br(CH₂)₃Cl, K₂CO₃, acetone, reflux, >90%; (**d**) bromine, ether, 0°C-rt, 50%.

4)

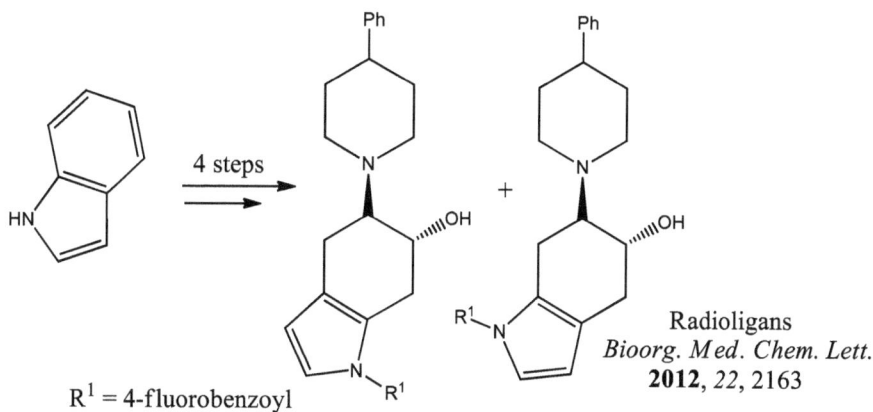

R¹ = 4-fluorobenzoyl

Radioligans
Bioorg. Med. Chem. Lett.
2012, *22*, 2163

(**a**) 4-fluorobenzoyl chloride/TEA/DMAP in CHCl₃, rt, 72 h; (**b**) 4-phenylpiperidine in ethanol, 80°C; 10 days; (**c**) Birch reduction, Li/NH₃/MeOH; (**d**) *m*-CPBA in CH₂Cl₂, rt, 5 days.

5)

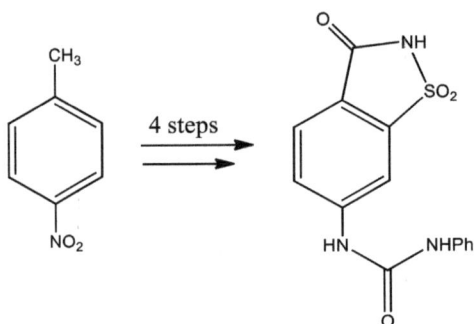

Saccharin analogues
Inhibitors of tyrosinase.
Bioorg. Med. Chem.
2012, *20*, 2811

(**a**) PhNCO, DMF, TEA, 12 h, rt; (**b**) CrO$_3$-H$_2$SO$_4$, rt; 24h; (**c**) H$_2$; Pd-C-EtOH; (**d**) NH$_3$, ether, 40°C, 2 h; (**e**) ClSO$_3$H, 60°C, 20h.

6)

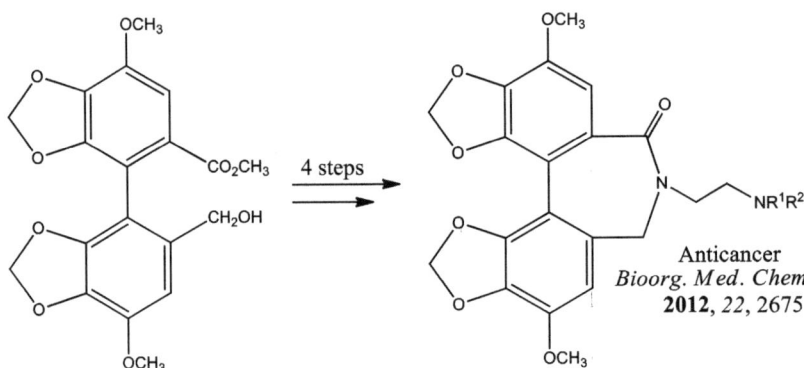

Anticancer
Bioorg. Med. Chem. Lett.
2012, *22*, 2675

(**a**) SOCl$_2$, 1H-benzotriazole, CH$_2$Cl$_2$, 0°C, 2 h, 98%; (**b**) different amines, KI, CH$_2$Cl$_2$, reflux, 2 h, 84–91%; (**c**) SOCl$_2$, CH$_2$Cl$_2$, 0°C, 2 h, 93%; (**d**) ethanolamine, Et$_3$N, KI, CH$_2$Cl$_2$, reflux, 3h, 86%.

7)

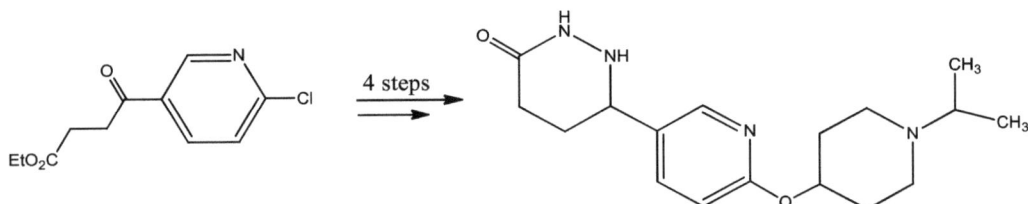

Histamine receptor reverse agonists.
Bioorg. Med. Chem. Lett.
2012, *22*, 1504

(**a**) acetone, NaCNBH₃, 80°C; (**b**) 4-hydroxy-1-Boc-piperidine, 1 mol/L *t*-BuOK, DMSO, 100°C; (**c**) NH₂NH₂.H₂O, EtOH, 80°C; (**d**) 4 mol/L HCl, dioxane, 50°C.

8)

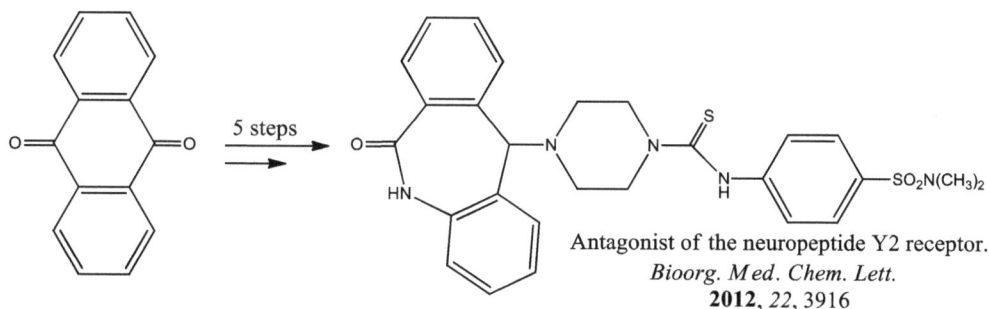

Antagonist of the neuropeptide Y2 receptor.
Bioorg. Med. Chem. Lett.
2012, *22*, 3916

(**a**) SOCl₂, CHCl₃, reflux, 4 h; (**b**) NaN₃, H₂SO₄, CHCl₃, reflux, 12 h; (**c**) NaBH₄, MeOH, 0°C at rt, 3 h; (**d**) 4-isothiocyanato-*N,N*-dimethylbenzenesulfonamide, CH₂Cl₂, rt, 3 h; (**e**) piperazine, *i*-Pr₂EtN, DMF, 145°C, microwaves, 1 h.

9)

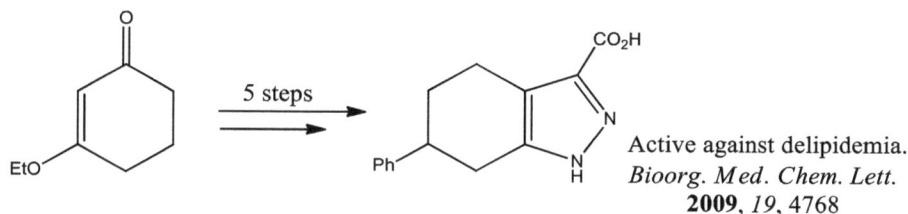

Active against delipidemia.
Bioorg. Med. Chem. Lett.
2009, *19*, 4768

(**a**) LiOH, MeOH/THF/H₂O; (**b**) LDA, THF, -78°C followed by diethyl oxalate; (**c**) PhMgCl, THF, -78°C at rt; (**d**) hydrazine (NH₂NH₂), AcOH:EtOH (1:10), 65°C; (**e**) Pd/C, H₂, MeOH.

10)

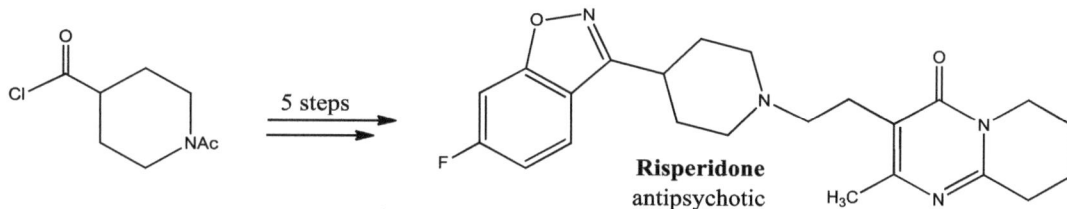

Risperidone
antipsychotic

(**a**) KOH, H₂O, reflux; (**b**) NH₂OH, EtOH, reflux; (**c**) 1,3-difluorobenzene, AlCl₃, DCM; (**d**) 6 mol/L HCl, reflux;

(**e**) Na₂CO₃, KI

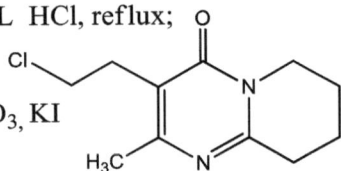

11)

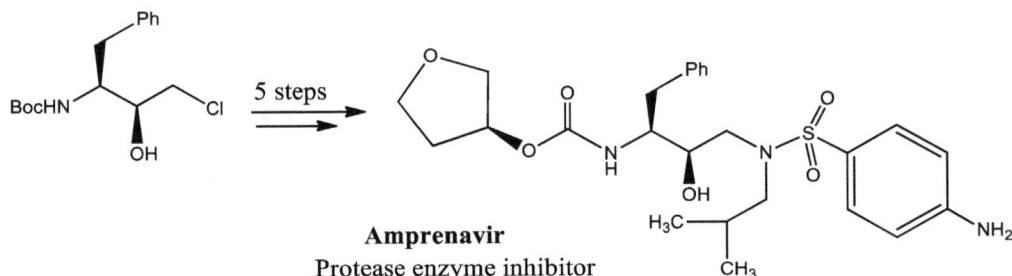

Amprenavir

Protease enzyme inhibitor
drug of the HIV virus.
Bioorg. Med. Chem. Lett.
2012, *22*, 1976

(**a**) 4-nitrobenzenesulfonyl chloride, NaHCO₃, DCM, H₂O; (**b**) isobutylamine, EtOH; (**c**) Pd–C, NH₂NH₂.H₂O, EtOH; (**d**) HCl, dioxane; (**e**) triphosgene or phosgene, TEA, (*S*)-(+)-3-hydroxytetrahydrofuran, DCM.

12)

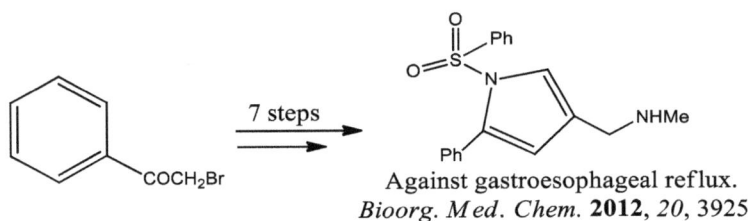

Against gastroesophageal reflux.
Bioorg. Med. Chem. **2012**, *20*, 3925

(**a**) MeNH₂, NaBH₃CN, THF, rt, 18 h; (**b**) HCl₍g₎–AcOEt, rt, 3 h (hint: in this step the formation of the chlorinated indole nucleus occurs at position 2, remembering that there are two more substitutions in the ring); (**c**) H₂, 10% Pd–C, EtOH. rt, 24 h; (**d**) NaH, PhSO₂Cl, THF, rt, 15 min; (**e**) DIBAL 1,5 mol/L in toluene, THF, −78 °C, 1 h; (**f**) ethyl cyanoacetate, K₂CO₃, acetone, rt, 18 h; (**g**) NMO, TPAP, molecular sieve 4 Å, MeCN, rt, 1.5 h.

13)

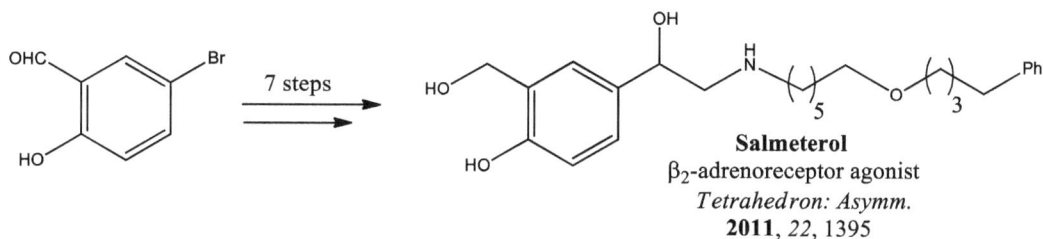

Salmeterol
β_2-adrenoreceptor agonist
Tetrahedron: Asymm.
2011, *22*, 1395

(a) $Br(CH_2)_6CH_2O(CH_2)_3CH_2Ph$; **(b)** CH_3NO_2, Et_3N, MeOH; **(c)** AcOH, H_2O, 70°C; **(d)** $NaBH_4$, THF; **(e)** *n*-BuLi, DMF, THF; **(f)** H_2/Pd-C, MeOH; **(g)** $Me_2C(OCH_3)_2$, TsOH, acetone.

14)

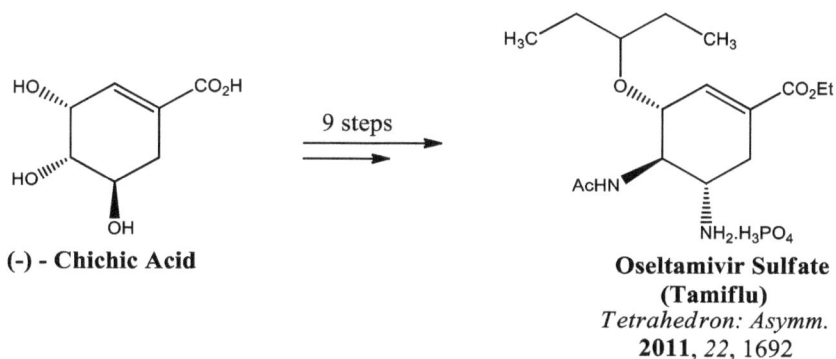

(-) - Chichic Acid

**Oseltamivir Sulfate
(Tamiflu)**
Tetrahedron: Asymm.
2011, *22*, 1692

(a) 2.0 eq. NaN_3 in DMF anhydrous, 90°C, 8 h; **(b)** 1.2 eq. Ac_2O, 2,0 eq. Et_3N in CH_2Cl_2, 0°C for 0.5 h; **(c)** 1.1 eq. Ph_3P, reflux for 8 h in THF/H_2O (10:1); 1.2 eq. H_3PO_4, 50°C for 2 h, then rt for 8 h in AcOEt/EtOH (2:1); **(d)** 1.5 eq. $BF_3.OEt_2$, -8°C for 1.5 h in 3-pentanol; **(e)** 2.5 eq. $SOCl_2$, 2.5 eq. Et_3N, 0-5°C for 2 h, and then rt for 12 h in AcOEt; **(f)** 1.1 eq. Ph_3P, 0.1 eq. diethylammonium 4-toluenesulfonate (Et$_2$NH-TsOH), rt for 1 h, and then 60°C for 2 h; **(g)** $SOCl_2$, EtOH, reflux, 3h; **(h)** 2.0 eq. NaN_3, reflux for 12 h in EtOH anhydrous; **(i)** 1.5 eq. MsCl, 1.2 eq. Et_3N, 0.1 eq. DMAP 0°C for 1 h in EtOAc.

15)

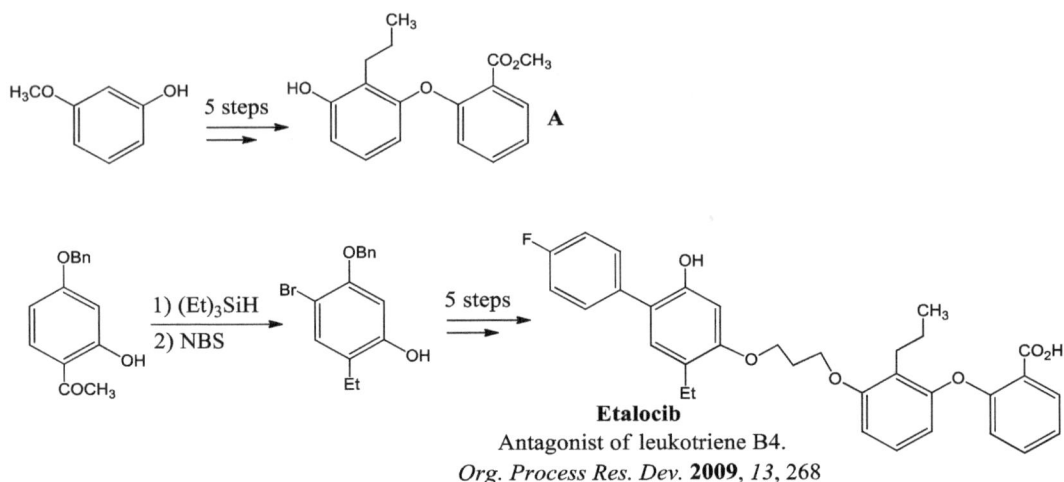

Etalocib
Antagonist of leukotriene B4.
Org. Process Res. Dev. **2009**, *13*, 268

(**a**) KI; (**b**) H₂ Pd-C; (**b**) 2.0 eq. *n*-BuLi, -78°C, followed by propyl iodide, and after completion of the reaction, the reaction medium is slightly acidic; (**c**) BBr₃; (**d**) *p*-B(OH)₂C₆H₄F, K₂CO₃, Pd(PPh₃)₄; (**e**) NaOH, the **A**, and after completion of the reaction, the reaction medium is slightly acidic; (**f**) 1-cyano-2-fluorobenzene, KF; (**g**) MeOH, H₂SO₄; (**h**) KOH, aq.; (**i**) BrCH₂CH₂CH₂Cl; (**j**) H₂ Pd-C.

16)

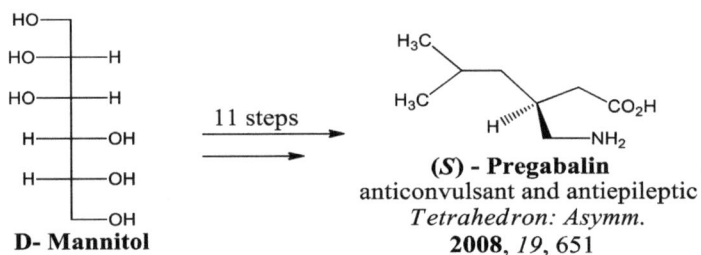

D- Mannitol

(*S*) - Pregabalin
anticonvulsant and antiepileptic
Tetrahedron: Asymm.
2008, *19*, 651

(**a**) H₂, HCl, 20% Pd(OH)₂/C, MeOH; (**b**) NH₄HCO₂, 20% Pd(OH)₂/C, MeOH; (**c**) 90% AcOH; (**d**) NaIO₄, THF-H₂O; (**e**) NaIO₄, THF-H₂O; (**f**) 1 mol/L LiOH, THF; (**g**) (EtO)₂POCHCO₂Et, CH₂Cl₂ ; (**h**) CH₃NO₂, THF; (**i**) (Boc)₂O, Et₃N, DMAP; (**j**) acetone H₂SO₄; (**k**) Ph₃P=C(CH₃)₂.

Answers

1) (**a**) bromination, Br₂, AcOH, rt; (**b**) Heck reaction with styrene; (**c**) nitromethane and base (NH₄OAc), 100°C, 3h.

2) (a) Thionyl chloride; **(b)** AlCl$_3$, PhCl; **(c)** 5-bromopyrimidine, *n*-BuLi, THF, -78°C.

3) (a) Br(CH$_2$)$_3$Cl, K$_2$CO$_3$, acetone, reflux,> 90%; **(b)** bromine, ether, 0°C - rt, 50%; **(c)** R$_2$R$_1$NH, K$_2$CO$_3$, MeCN, 80°C, > 90%; **(d)** (*R*)-2-methylpyrrolidine.HCl, NaI, 2-butanone, reflux, 30%.

4) (a) Reduction of Birch, Li/NH$_3$/MeOH; **(b)** 4-fluorobenzoyl chloride/TEA/DMAP in CHCl 3, rt, 72 h; **(c)** *m*-CPBA in CH$_2$Cl$_2$, rt, 5 days; **(d)** 4-phenylpiperidine in ethanol, 80°C; 10 days.

5) (a) ClSO$_3$H, 60°C, 20h; **(b)** NH$_3$, ether, 40°C, 2 h; **(c)** CrO$_3$-H$_2$SO$_4$, rt, 24h; **(d)** H$_2$; Pd-C-EtOH; **(e)** PhNCO, DMF, TEA, 12 h, rt.

6) (a) SOCl$_2$, 1*H*-benzotriazole, CH$_2$Cl$_2$, 0°C, 2 h, 98%; **(b)** ethanolamine, Et$_3$N, KI, CH$_2$Cl$_2$, reflux, 3 h, 86%; **(c)** SOCl$_2$, CH$_2$Cl$_2$, 0°C, 2 h, 93%; **(d)** different amines, KI, CH$_2$Cl$_2$, reflux, 2 h, 84-91%.

7) (a) NH$_2$NH$_2$·2H$_2$O, EtOH, 80°C; **(b)** 4-hydroxy-1-Boc-piperidine, 1 mol/L *t*-BuOK, DMSO, 100°C; **(c)** 4 mol/L HCl, dioxane, 50°C; **(d)** acetone, NaCNBH$_3$, 80°C.

8) (a) NaN$_3$, H$_2$SO$_4$, CHCl$_3$, reflux, 12 h; **(b)** NaBH$_4$, MeOH, 0°C at rt, 3 h; **(c)** SOCl$_2$, CHCl$_3$, reflux, 4 h; **(d)** piperazine, *i*Pr$_2$EtN, DMF, 145°C, microwaves, 1 h; **(e)** 4-isothiocyanato-*N*,*N*-dimethylbenzenesulfonamide, CH$_2$Cl$_2$, rt, 3 h.

9) (a) PhMgCl, THF, -78°C at rt; **(b)** Pd/C, H$_2$, MeOH; **(c)** LDA, THF, -78°C followed by ethyl oxalate; **(d)** hydrazine (NH$_2$NH$_2$), AcOH:EtOH (1:10), 65°C; **(e)** LiOH, MeOH/THF/H$_2$O.

10)

(a) 1,3-difluorobenzene, AlCl$_3$, DCM; **(b)** 6 mol/L HCl, reflux; **(c)** NH$_2$OH, EtOH, reflux;

(d) KOH, H$_2$O, reflux; **(e)**

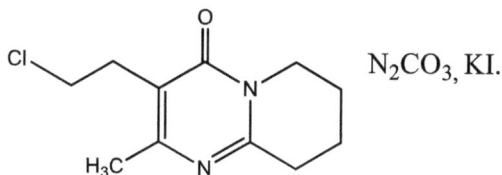

N$_2$CO$_3$, KI.

(a)

(b) Cleavage of the acetyl group. (c)

(d)

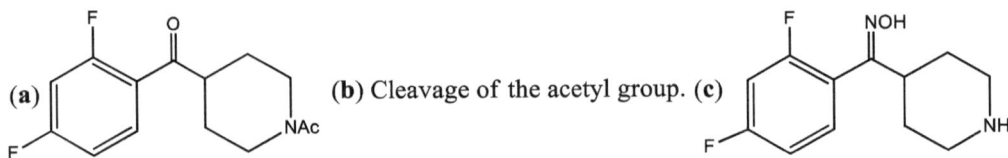

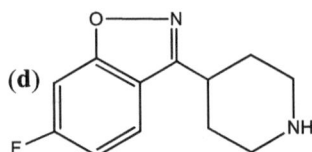

(e) Reaction of coupling, drug formation risperidone.

11) **(a)** isobutylamine, EtOH; **(b)** 4-nitrobenzenesulfonyl chloride, NaHCO$_3$, DCM, H$_2$O; **(c)** HCl, dioxane; **(d)** phosgene, TEA, (*S*)-(+)- 3-hydroxytetrahydrofuran, DCM; **(e)** Pd–C, NH$_2$NH$_2$.H$_2$O, EtOH.

12) **(a)** ethyl cyanoacetate, K$_2$CO$_3$, acetone, rt, 18 h; **(b)** HCl$_{(g)}$–AcOEt, rt, 3 h; **(c)** H$_2$, 10% Pd–C, EtOH. rt, 24 h; **(d)** NaH, PhSO$_2$Cl, THF, rt, 15 min; **(e)** 1.5 mol/L DIBAL in toluene, THF, −78 °C, 1 h; **(f)** NMO, TPAP, molecular sieve 4 Å, MeCN, rt, 1.5 h; **(g)** MeNH$_2$, NaBH$_3$CN, THF, rt, 18 h.

RUTHENIUM AND ITS COMPLEXES AS OXIDANTS

Ruthenium and its complexes are important reagents in modern organic synthesis capable of oxidizing a variety of functional groups catalytically, selectively and under mild conditions. A widely used reagent is TPAP (tetrapropylammonium perruthenate) in the presence of NMO (*N*-methylmorpholine *N*-oxide), while latter reagent is used as a co-oxidant so that ruthenium can be reoxidized and used under catalytic conditions (Scheme 1).

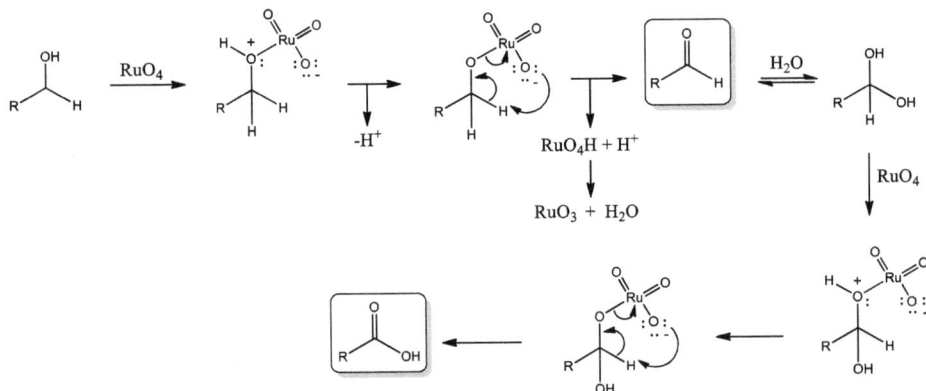

RuO$_4$ Regeneration

NHMOH
N-methylmorpholine *N*-oxide
Co-oxidant

N-methylmorpholine

Ru (VI) Ru (VIII)

Oxidation of ruthenium by the co-oxidant responsible for its reuse.

N-methylmorpholine
+
RuO$_4$

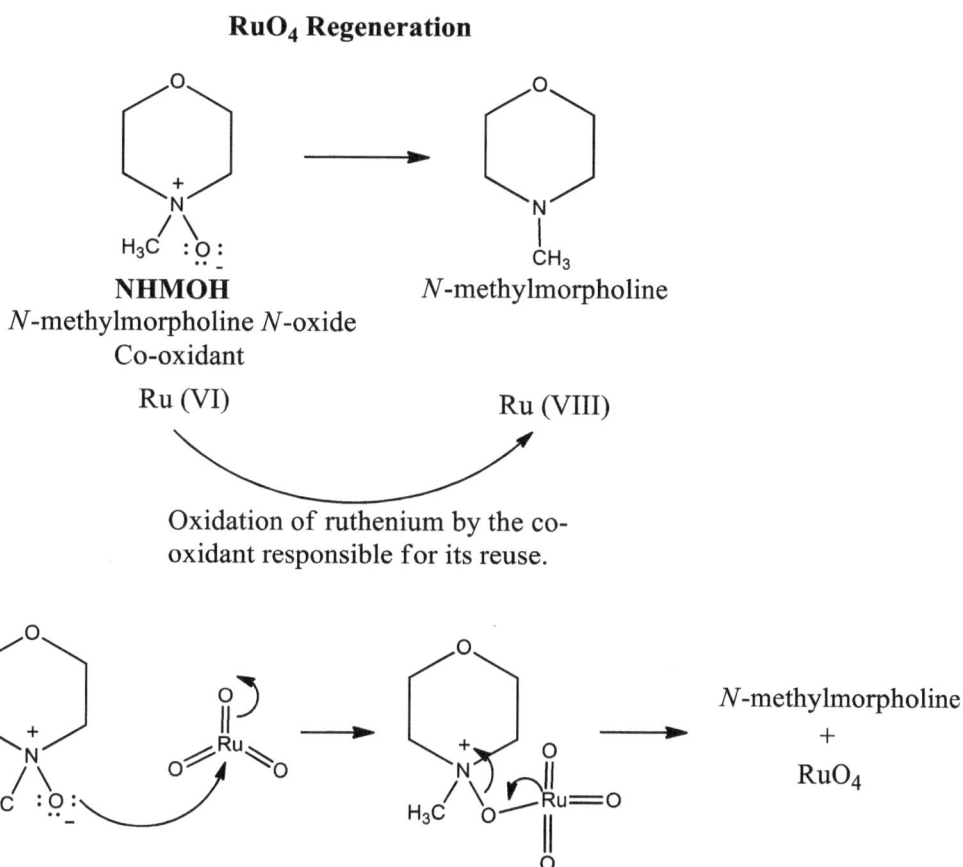

Reference: *Org. Lett.* **2011**, *13*, 4164.

Scheme 1. Mechanism proposed for the oxidation of alcohols in the presence of TPAP-NMO.

13) (a) NaBH$_4$, THF; **(b)** Me$_2$C(OCH$_3$)$_2$, TsOH, acetone; **(c)** *n*-BuLi, DMF, THF; **(d)** CH$_3$NO$_2$, Et$_3$N, MeOH; **(e)** H$_2$/Pd-C, MeOH; **(f)** Br(CH$_2$)$_6$CH$_2$O(CH$_2$)$_3$CH$_2$Ph; **(g)** AcOH, H$_2$O, 70°C.

14) (a) SOCl$_2$, EtOH, reflux, 3 h; **(b)** 2.5 eq. SOCl$_2$, 2.5 eq. Et$_3$N, 0-5°C por 2 h, and then rt for 12 h in AcOEt; **(c)** 2.0 eq. NaN$_3$, reflux for 12 h in anhydrous ethanol; **(d)** 1.1 eq. Ph$_3$P, 0.1 eq. diethylammonium *p*-toluenesulfonate (Et$_2$NH-TsOH), rt for 1 h, and then 60°C for 2 h; **(e)** 1.2 eq. Ac$_2$O, 2.0 eq. Et$_3$N, 0°C for 0.5 h in CH$_2$Cl$_2$. **(f)** 1.5 eq. BF$_3$.OEt$_2$, -8°C for 1.5 h in 3-pentanol; **(g)** 1.5 eq. MsCl, 1.2 eq. Et$_3$N, 0.1 eq. DMAP 0°C for 1 h in EtOAc; **(h)** 2.0 eq. NaN$_3$, 90°C for 8 h anhydrous DMF; (**i**) 1.1 eq. Ph$_3$P, reflux for 8 h in THF/H$_2$O (10:1); 1.2 eq. H$_3$PO$_4$, 50°C for 2 h, then rt for 8 h in AcOEt/EtOH (2:1).

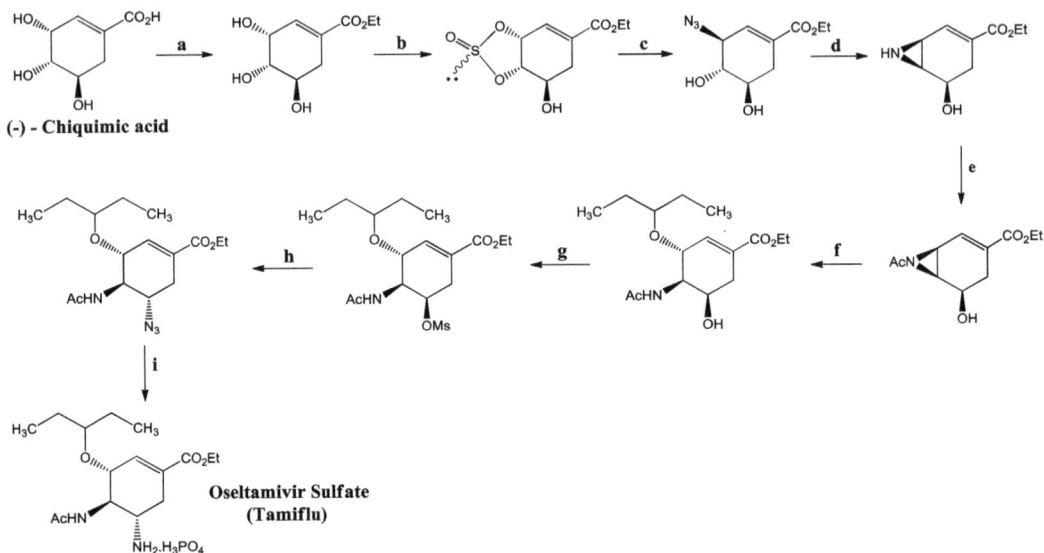

15) (**a**) *n*-BuLi, -78°C, followed by propyl iodide, and after the completion of the reaction the medium was slightly acidified (ortho metalation reaction); (**b**) 1-cyano-2-fluorobenzene, KF; (**c**) KOH, heat; (**d**) MeOH, H$_2$SO$_4$; (**e**) BBr$_3$; (**f**) BrCH$_2$CH$_2$CH$_2$Cl; (**g**) *p*-B(OH)$_2$C$_6$H$_4$F, K$_2$CO$_3$, Pd(PPh$_3$)$_4$ (Suzuki coupling); (**h**) KI reaction of chlorine exchange by iodine to facilitate the reaction of SN2; (**i**) NaOH, then **A**, and after the completion of the reaction the medium was slightly acidified (coupling and saponification reaction); (**j**) H$_2$ Pd-C (deprotection of the benzyl group). It is worth mentioning that the halogenolysis reaction in the presence of palladium only occurs under harsh conditions due to the strength of the C-F bond (about 110 kcal / mol). **16**) (**a**) acetone H$_2$SO$_4$; (**b**) NaIO$_4$, THF-H$_2$O; (**c**) (EtO)$_2$POCHCO$_2$Et, CH$_2$Cl$_2$; (**d**) CH$_3$NO$_2$, THF; (**e**) NH$_4$HCO$_2$, 20% Pd(OH)$_2$/C, MeOH; (**f**) (Boc)$_2$O, Et$_3$N, DMAP; (**g**) 90% AcOH; (**h**) NaIO$_4$, THF-H$_2$O; (**i**) Ph$_3$P=C(CH$_3$)$_2$; (**j**) 1 mol/L LiOH, THF; (**k**) H$_2$, HCl, 20% Pd(OH)$_2$/C, MeOH.

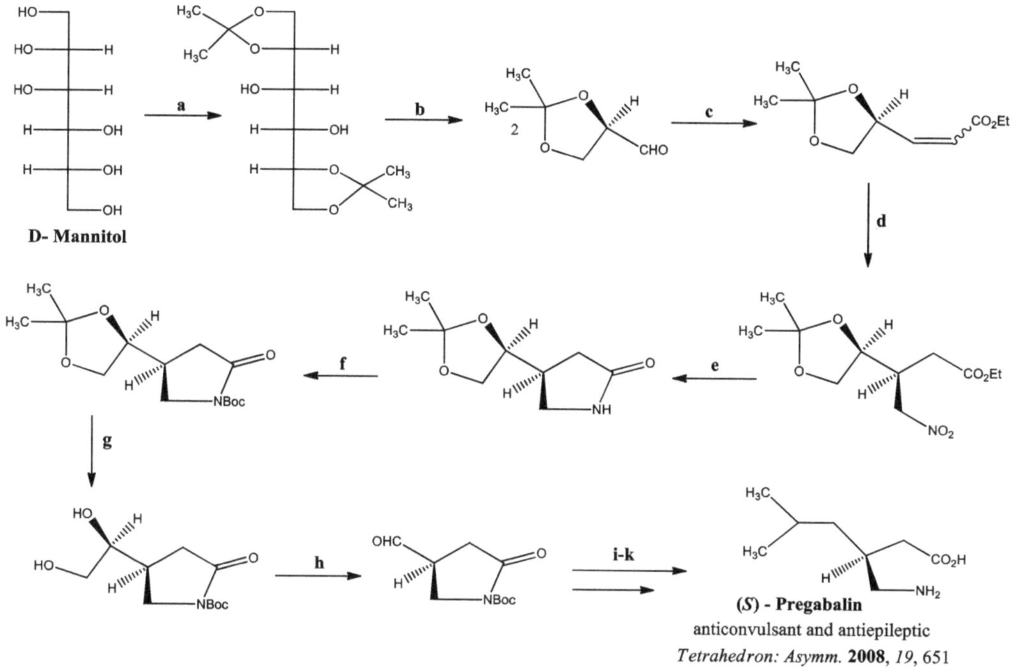

D- Mannitol

(S) - Pregabalin
anticonvulsant and antiepileptic
Tetrahedron: Asymm. **2008**, *19*, 651

MECHANISM OF WITTIG-HORNER REACTION

This reaction is a modification of the Wittig reaction known as the Wittig-Horner reaction, in which the ylide is obtained using a phosphonate (Scheme **2**).

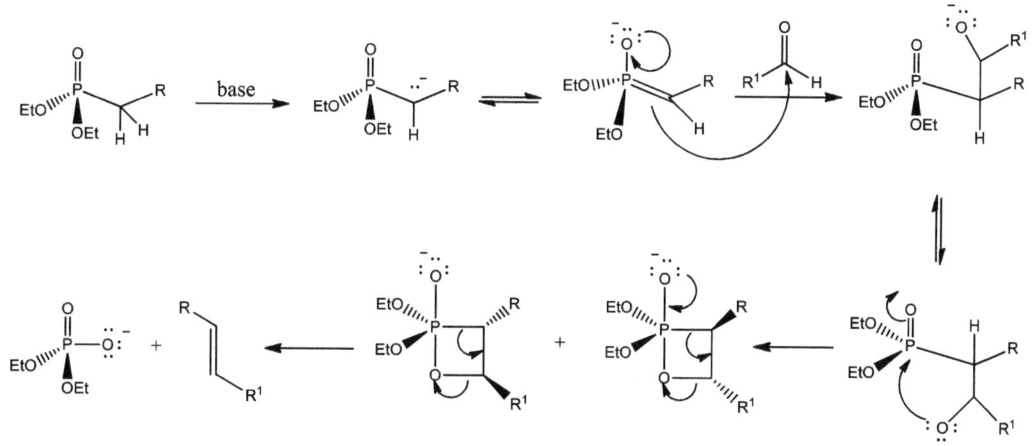

Scheme 2. Proposed mechanism of the Wittig-Horner reaction.

EXERCISE 6 - IDENTIFYING REAGENTS

The next exercise is to identify reagents used for the preparation of a specific target molecule. This type of training is essential because it allows the reader to contact different types of reagents and chemical transformations, as well as to use them in a sequentially correct way.

Example

What are the five steps necessary for the synthesis of the drug pyrovalerone? This psychoactive drug is a stimulant used in chronic fatigue, lethargy as well as loss of appetite, one of its side effects.

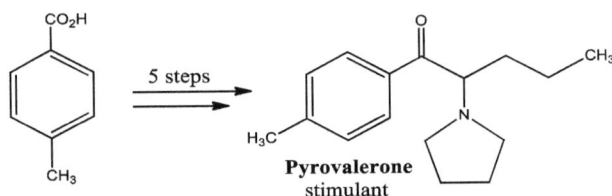

Pyrovalerone
stimulant

Answer

As we already mentioned, retrosynthetic analysis is also crucial in this type of exercise. We can extract different information, such as synthons and their respective reagents, as well as the order of their uses. In this context, we can see from the structure of the drug in question that there is a substitution at the α-carbonyl (pyrrolidine) position, as well as carbon-carbon bond formation (CO-C). Thus, the pyrrolidine reagent, through a leaving group, bromine, for example, is easily inserted in the α-carbonyl position. The formation of the CO-C bond can be effected using the Grignard reagent. The retrosynthetic analysis also guides us about the order of the reagents to be employed. For example, for the introduction of the pyrrolidine group in the α-carbonyl position, the formation of the CO-C bond is first necessary. Thus, the conversion of the carboxylic acid into acid chloride is first required, followed by the Grignard reagent, base, bromine, and finally, the use of the pyrrolidine.

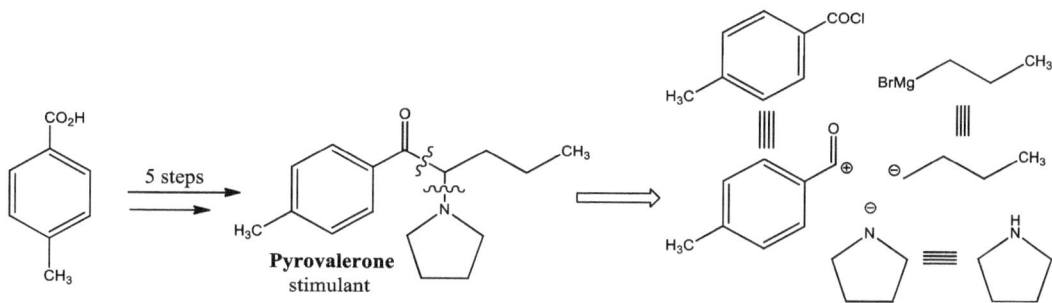

Pyrovalerone
stimulant

(**a**) thionyl chloride; (**b**) butylmagnesium chloride (Grignard reagent); (**c**) base; bromo; (**d**) pyrrolidine.

EXERCISES

1)

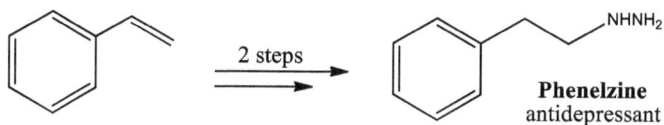

Phenelzine
antidepressant

2)

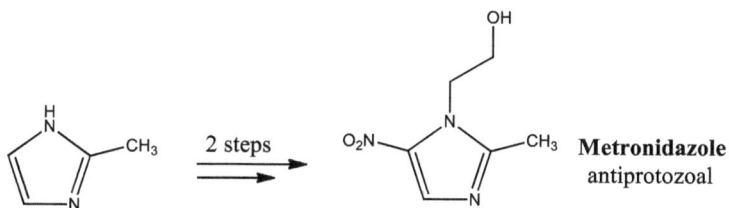

Metronidazole
antiprotozoal

3)

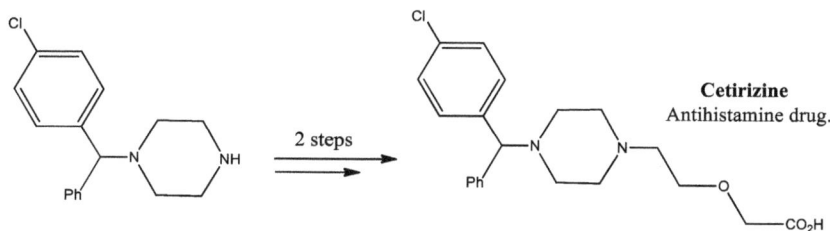

Cetirizine
Antihistamine drug.

4)

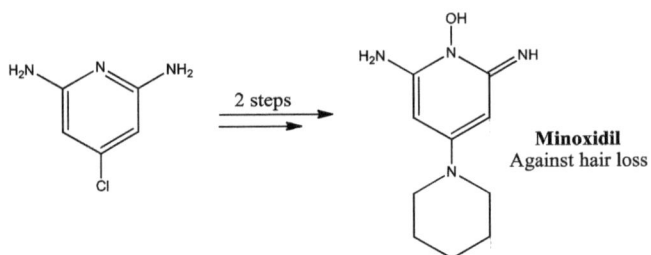

Minoxidil
Against hair loss

2 steps

5)

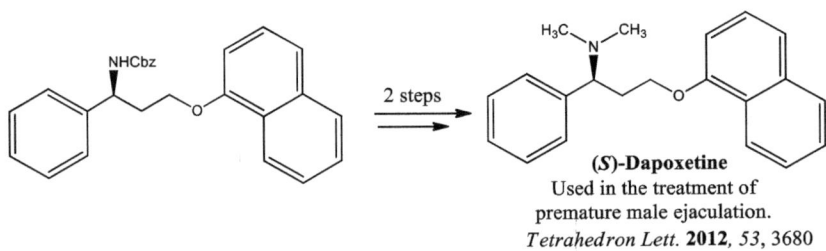

2 steps

(S)-Dapoxetine
Used in the treatment of
premature male ejaculation.
Tetrahedron Lett. **2012**, *53*, 3680

6)

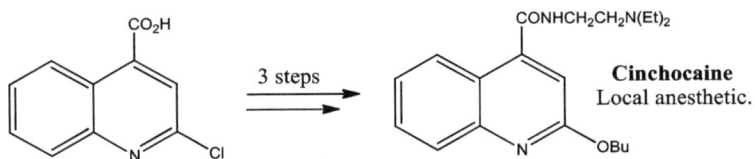

3 steps

Cinchocaine
Local anesthetic.

7)

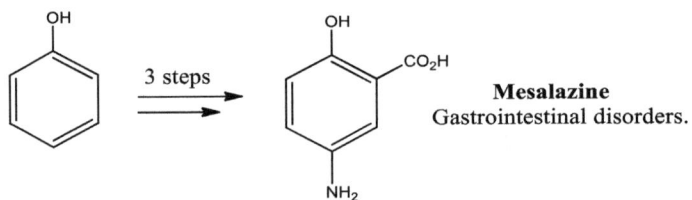

3 steps

Mesalazine
Gastrointestinal disorders.

8)

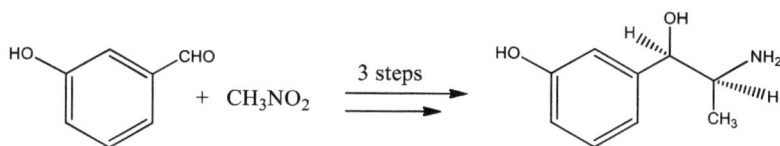

Metaraminol
Employed in the prevention and treatment
of hypotension, especially that caused by
anesthesia.

9)

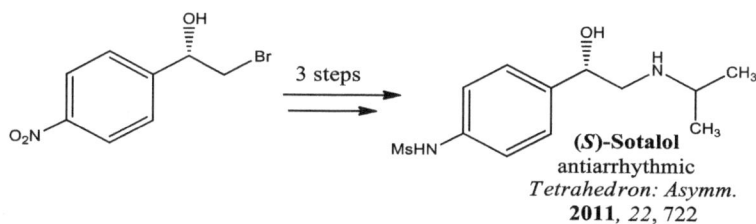

(S)-Sotalol
antiarrhythmic
Tetrahedron: Asymm.
2011, *22*, 722

10)

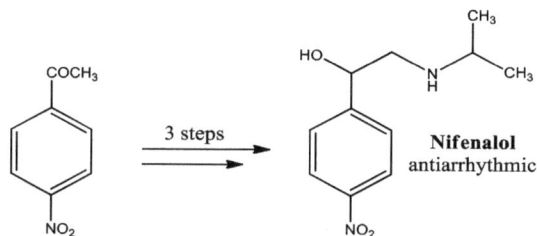

Nifenalol
antiarrhythmic

11)

Phenylacetone   3 steps

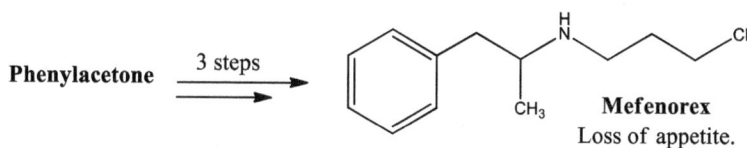

Mefenorex
Loss of appetite.

12)

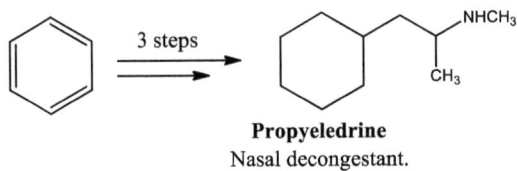

Propyeledrine
Nasal decongestant.

13)

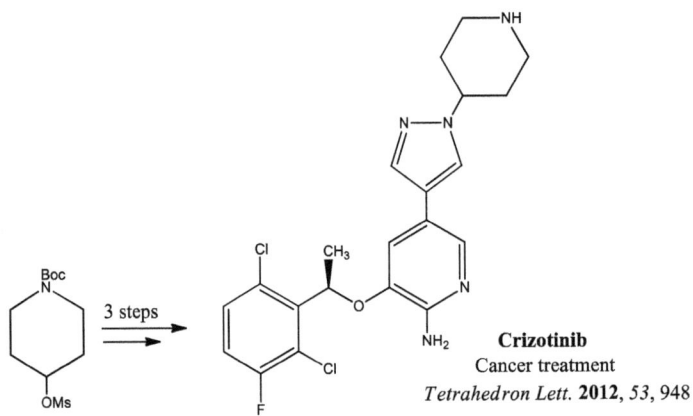

Crizotinib
Cancer treatment
Tetrahedron Lett. **2012**, *53*, 948

14)

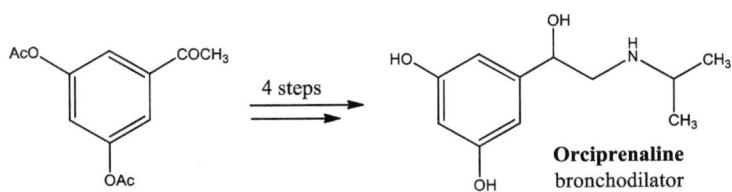

Orciprenaline
bronchodilator

15)

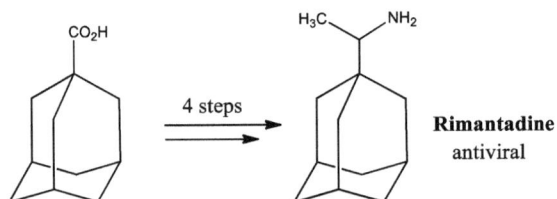

Rimantadine
antiviral

16)

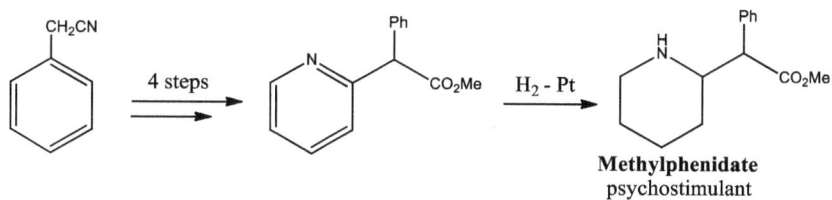

Methylphenidate
psychostimulant

17)

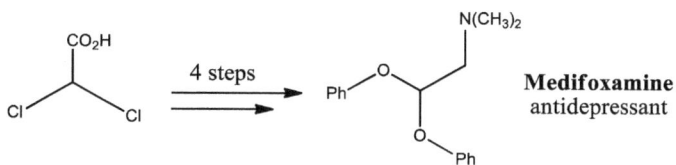

Medifoxamine
antidepressant

18)

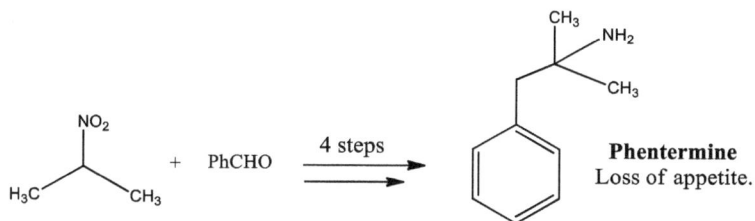

Phentermine
Loss of appetite.

19)

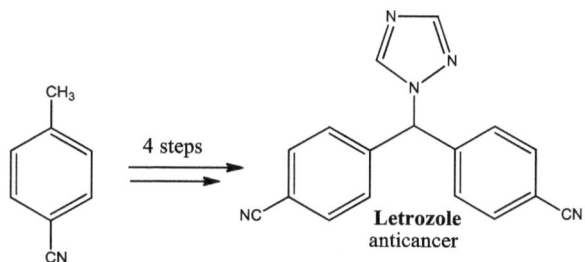

Letrozole
anticancer

20)

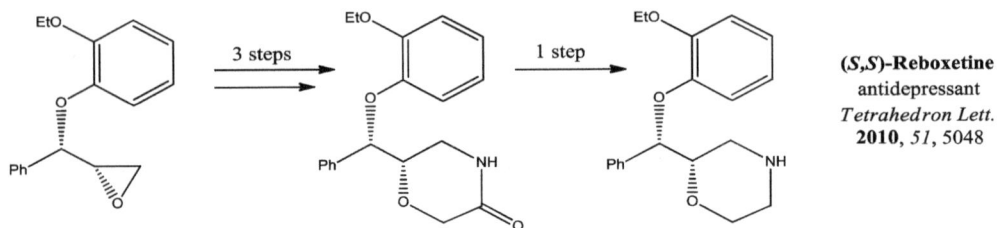

(*S,S*)-**Reboxetine**
antidepressant
Tetrahedron Lett.
2010, *51*, 5048

21)

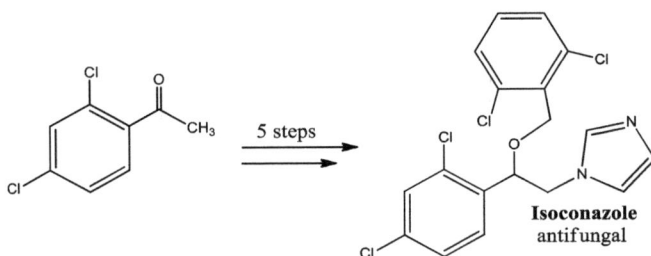

Isoconazole
antifungal

22)

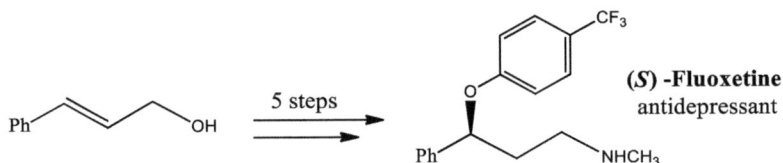

(*S*) -**Fluoxetine**
antidepressant

23)

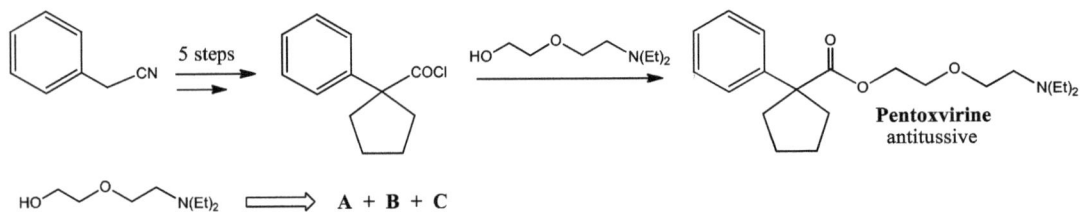

Pentoxvirine
antitussive

$HO\!\!\sim\!\!\sim\!\!O\!\!\sim\!\!N(Et)_2 \implies A + B + C$

Answers

1) (**a**) HBr, hydrogen peroxide; (**b**) hydrazine.

2) (**a**) nitration; (**b**) epoxide.

3) (**a**) epoxide or 1-chloroethanol; (**b**) chloroacetic acid.

4) (**a**) metachloroperbenzoic acid; (**b**) piperidine.

5) (**a**) H_2, 10% Pd/C, MeOH/EtOAc (1:1); (**b**) $2HCO_2H$, 30% 2HCOH, reflux, 8 h.

6) (**a**) thionyl chloride in ethanol or ethanol in an acid medium; (**b**) $NH_2CH_2CH_2N(Et)_2$, (**c**) BuO^-Na^+.

7) (**a**) CO_2 - pressure; (**b**) nitration; (**c**) reduction of the nitro group.

8) (**a**) base; (**b**) reduction of the nitro group; (**c**) separation of diastereoisomers.

9) (**a**) H_2-Pt/c has to be this metal since if we use palladium, we have halogenolysis; (**b**) MsCl (1.0 equivalent to not mesylate the hydroxyl group); (**c**) isopropylamine. An alternative to using palladium instead of platinum is to begin synthesis first with the introduction of the isopropylamine group.

10) (**a**) bromination; (**b**) reduction of the carbonyl group; (**c**) isopropylamine.

11) (**a**) 3-amino-1-propanol; (**b**) hydrogenation (reductive amination); (**c**) thionyl chloride.

12) (**a**) 1-chloroacetone, $AlCl_3$; (**b**) methylamine; (**c**) H_2-PtO_2 pressure.

13)

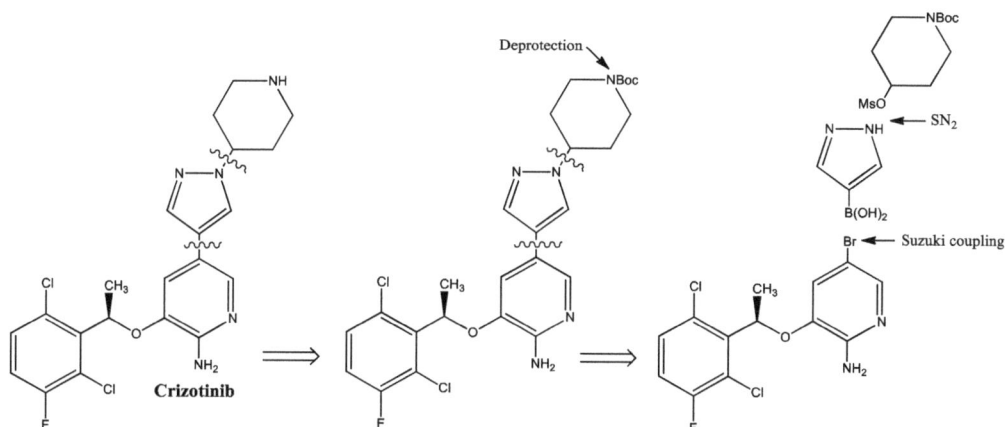

14) (**a**) bromo; (**b**) isopropylamine; (**c**) deprotection of the acetyl groups (acid medium HCl, or basic NaOH); (**d**) reduction of the carbonyl (sodium borohydride).

15) (**a**) thionyl chloride; (**b**) methyl lithium; (**c**) ammonia; (**d**) hydrogenation (reductive amination).

16) (**a**) base; (**b**) 2-chloropyiridine; (**c**) H_2SO_4; (**d**) MeOH, HCl.

17) (**a**) phenol (2 moles); (**b**) thionyl chloride; (**c**) dimethylamine; (**d**) lithium aluminum hydride.

18) (**a**) base; (**b**) H_2-Raney nickel; (**c**) thionyl chloride (conversion of hydroxyl to chlorine); (**d**) H_2-Pd-$CaCO_3$ (halogenolysis reaction).

19) (**a**) bromine (light); (**b**) 1,2,4-triazole; (**c**) base; (**d**) 1-cyano-4-fluorobenzene.

20) (**a**) NH_4OH; (**b**) chloroacetyl chloride; (**c**) KOt-Bu; (**d**) $LiAlH_4$.

21) (**a**) bromine, methanol; (**b**) imidazole, methanol; (**c**) $NaBH_4$; (**d**) NaH; (**e**) 2,6-dichlorobenzyl chloride.

22) (**a**) asymmetric epoxidation of Sharples: Ti(Oi-Pr)$_4$, D-(-)-diisopropyl tartrate, t-BuOOH; (**b**) the epoxide opening with Red-Al [bis (2-methoxyethoxy) aluminum NaAlH$_2$(OC$_2$H$_4$OCH$_3$)$_2$] or LiAlH$_4$; (**c**) MsCl regioselective transformation of the primary hydroxyl in a good leaving group; (**d**) MeNH$_2$; (**e**) NaH, 1-fluoro-4-trifluoromethyl-benzene.

23) (**a**) $NaNH_2$; (**b**) 1,4-dibromobutane; (**c**) $NaNH_2$; (**d**) strong acid medium HCl; (**e**) thionyl chloride.

A = ethylene glycol; **B** = 1-bromo-2-chloroethane; **C** = diethylamine.

CONCLUSION OF THE EXERCISES 5 AND 6 (ORGANIZING SYNTHETIC ROUTES AND IDENTIFYING REAGENTS)

After the exercises on retrosynthetic analysis, the exercises above were based on other essential aspects of a particular synthetic route. It is worth mentioning that the retrosynthetic analysis has to be allied to solid chemical knowledge. Therefore, the identification of reactions, reagents, intermediates, and chemical transformations are essential. For example, Exercise **5** proposes different syntheses with the reagents arranged incorrectly, and it is up to the reader to order them. Exercise **6** aims at identifying the reagents used and their correct order of use in the preparation of a given drug (target molecule). These exercises are intended to make the reader familiar with these themes.

CHAPTER 3

Identifying Drug Structure and Completing the Gaps

Abstract: In this chapter, the exercises below are also based on two themes. The first has an objective of the identification of the structure of the drugs having information of the starting materials and reagents used. This activity also enables the reader to become familiar with different classical chemical transformations used in medicinal chemistry and with varying types of reagents, reaction conditions, and necessary intermediates. In the second theme, the exercises are based on a series of gaps in the synthesis of different drugs. Therefore, it is up to the reader to complete them. This type of exercise is a combination of the themes previously presented. It is necessary to identify the reagents' use, the reaction conditions employed, and the intermediates formed, besides the use of the concept of retrosynthesis.

Keywords: Bioactive compounds, Chemical transformations, Drugs, Exercises, Medicinal chemistry, Molecules, Reaction conditions, Reagents, Retrosynthetic analysis, Substances, Synthesis, Synthons.

EXERCISE 7 - IDENTIFYING DRUG STRUCTURE

The exercises below have an objective of the identification of the structure of the drugs having information of the starting materials and reagents used.

Examples

a) What is the structure of the drug lubiprostone after heating the intermediate below?

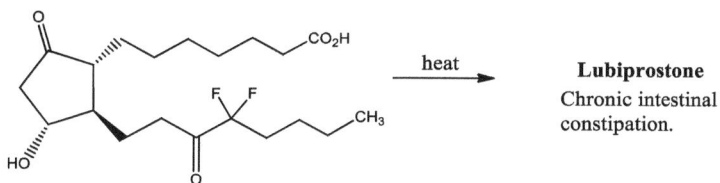

Lubiprostone
Chronic intestinal constipation.

Marcus Vinícius Nora de Souza

b) What is the structure of the drug below? Can step number two be performed with heating and excess ammonia? Justify.

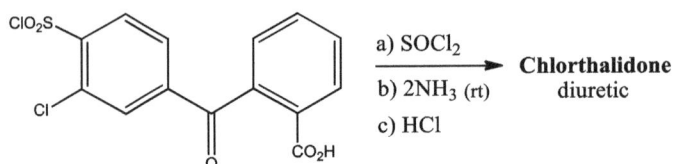

Answers

a) We can observe that an intramolecular cyclization occurs between the hydroxyl and the carbonyl, with the formation of a six-membered ring (hemiacetal - tautomerism chain-ring).

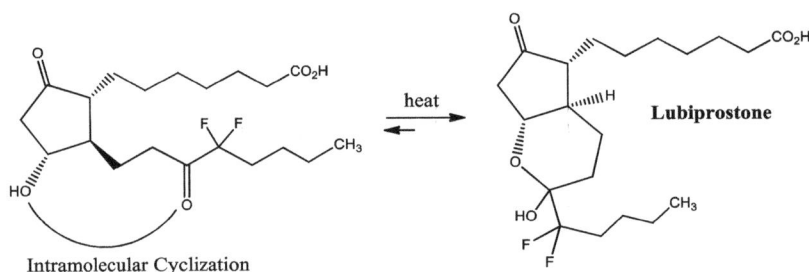

Intramolecular Cyclization

b) The first two steps are easy because they are classical and widely used transformations, thionyl chloride followed by ammonia, which is capable of converting the carboxylic acid and sulfonyl chloride into amide and sulfonamide, respectively. The third step requires a little more attention since an intramolecular cyclization occurs in an acid medium, with the formation of the indole nucleus. The reaction using ammonia can not be done with an excess of this reagent and heating, because an aromatic nucleophilic substitution reaction between the chlorine atom and the ammonia could occur.

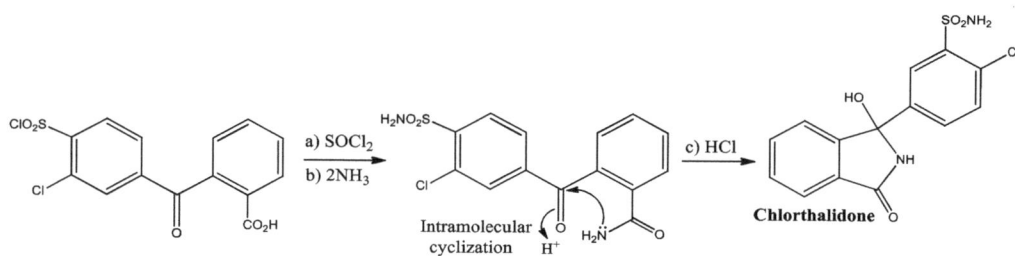

Chlorthalidone

EXERCISES

1)

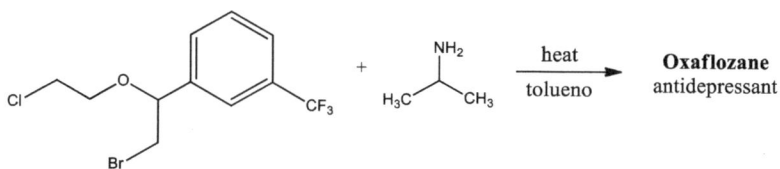

Oxaflozane
antidepressant

2)

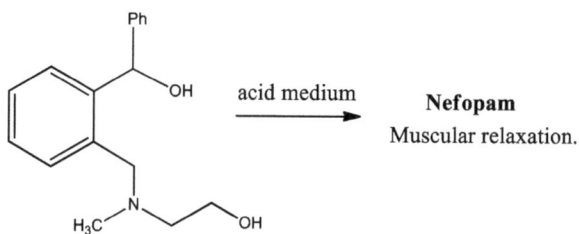

Nefopam
Muscular relaxation.

3)

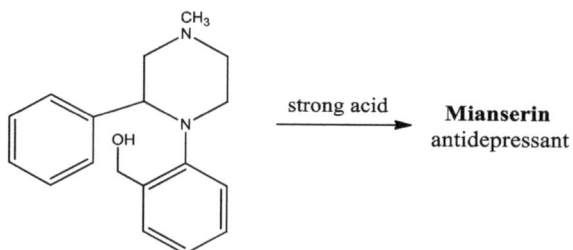

strong acid → **Mianserin**
antidepressant

4)

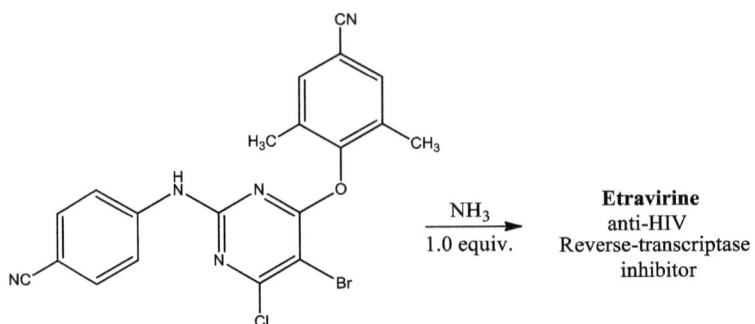

$$\xrightarrow[\text{1.0 equiv.}]{NH_3}$$

Etravirine
anti-HIV
Reverse-transcriptase
inhibitor

5) Tip: formation of a bicyclic system.

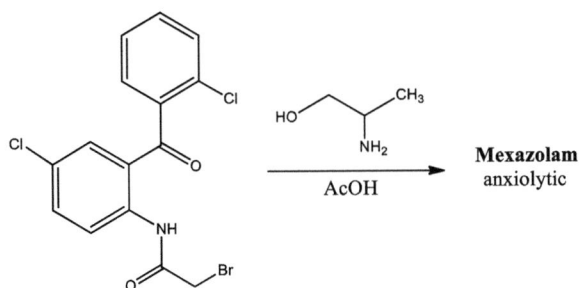

$$\xrightarrow{\text{AcOH}}$$

Mexazolam
anxiolytic

6)

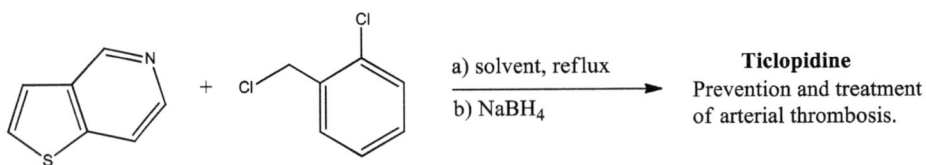

a) solvent, reflux

b) NaBH$_4$

Ticlopidine
Prevention and treatment
of arterial thrombosis.

7)

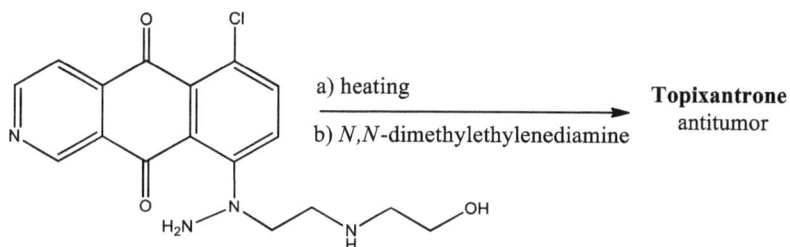

a) heating

b) *N,N*-dimethylethylenediamine

Topixantrone
antitumor

8)

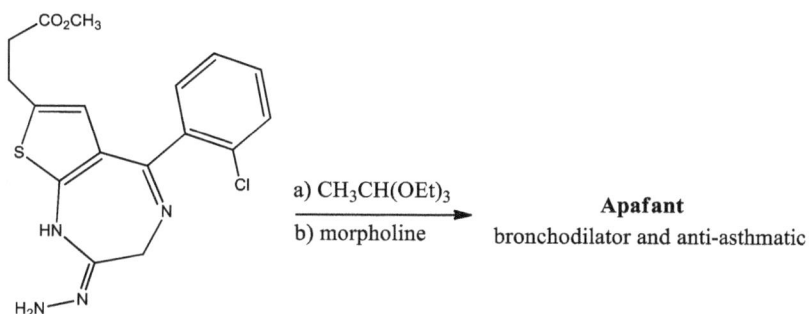

a) CH₃CH(OEt)₃

b) morpholine

Apafant
bronchodilator and anti-asthmatic

9)

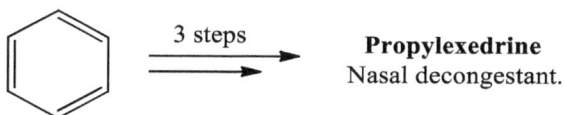

3 steps

Propylexedrine
Nasal decongestant.

(**a**) 1-chloroacetone, AlCl₃; (**b**) methylamine; (**c**) H₂-PtO₂ pressure.

10)

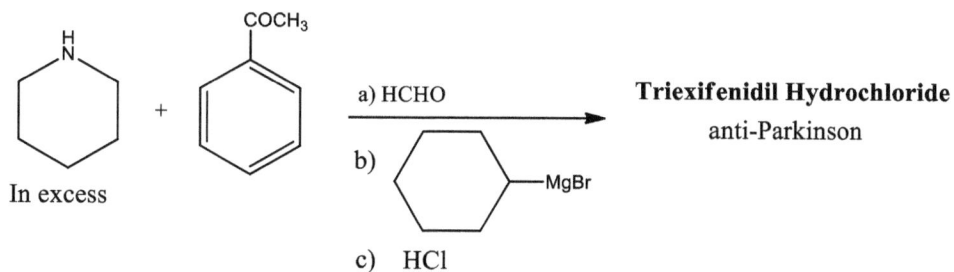

In excess   +

a) HCHO

b)   ⬡—MgBr

c)   HCl

Triexifenidil Hydrochloride
anti-Parkinson

11)

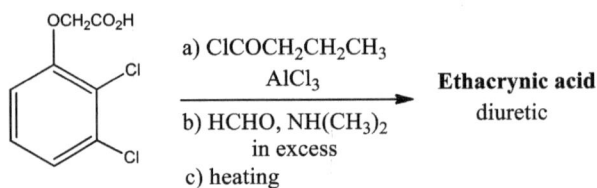

a) ClCOCH$_2$CH$_2$CH$_3$
   AlCl$_3$

b) HCHO, NH(CH$_3$)$_2$
   in excess

c) heating

Ethacrynic acid
diuretic

12)

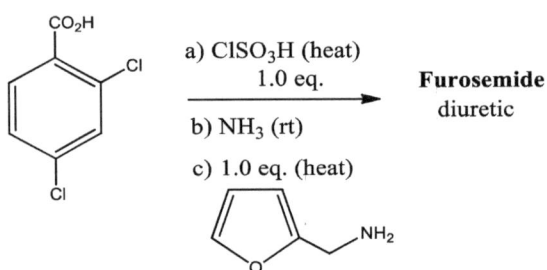

a) ClSO$_3$H (heat)
   1.0 eq.

b) NH$_3$ (rt)

c) 1.0 eq. (heat)

Furosemide
diuretic

13)

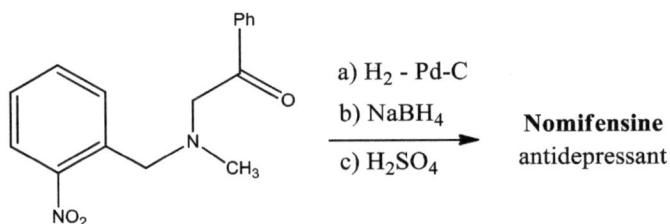

a) H$_2$ - Pd-C

b) NaBH$_4$

c) H$_2$SO$_4$

Nomifensine
antidepressant

14)

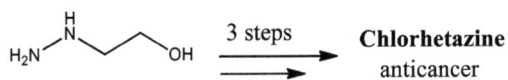

3 steps

Chlorhetazine
anticancer

(**a**) 3.0 eq. MsCl; (**b**) LiCl; (**c**) CH$_3$NCO.

(**a**) 3.0 eq. MsCl; (**b**) LiCl; (**c**) CH$_3$NCO.

15)

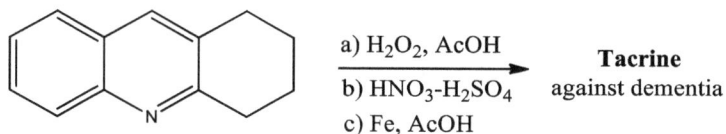

a) H_2O_2, AcOH

b) HNO_3-H_2SO_4

c) Fe, AcOH

Tacrine
against dementia

16)

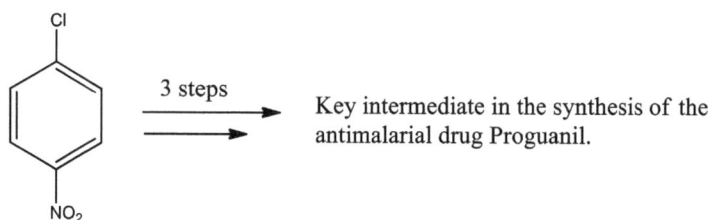

3 steps

Key intermediate in the synthesis of the antimalarial drug Proguanil.

(**a**) Fe, AcOH; (**b**) CNBr; (**c**) NH_3.

17)

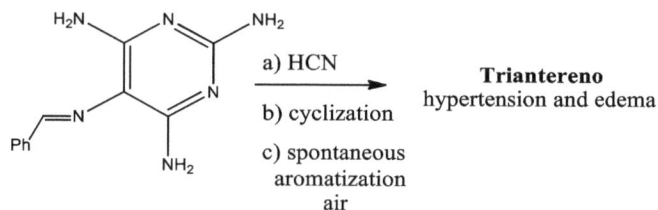

a) HCN

b) cyclization

c) spontaneous aromatization air

Triantereno
hypertension and edema

18)

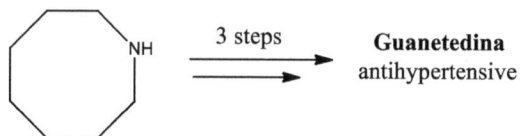

3 steps

Guanetedina
antihypertensive

(**a**) chloroacetonitrile; (**b**) hydrogenation; (**c**) $HN=C(SCH_3)(NH_2)$, heating (reaction occurs with release of methanethiol).

19)

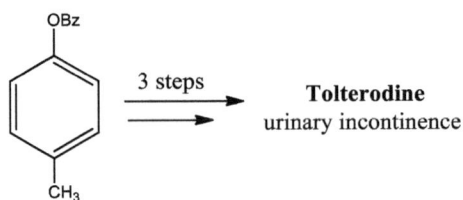

(a) AlCl$_3$; **(b)** LiCH$_2$CH$_2$N(i-Pr)$_2$; **(c)** H$_2$.

20)

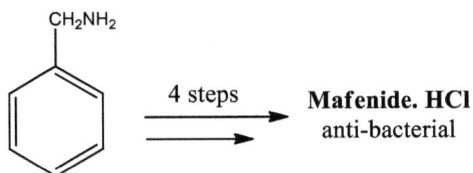

(a) Ac$_2$O; **(b)** ClSO$_2$OH; **(c)** NH$_3$; **(d)** 2.0 eq. HCl 2.0 mol/L, heat.

21)

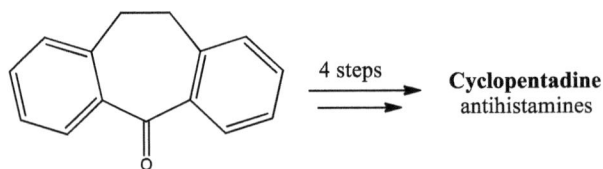

(a) NBS; **(b)** Et$_3$N; **(c)** 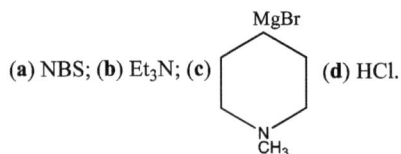 **(d)** HCl.

22)

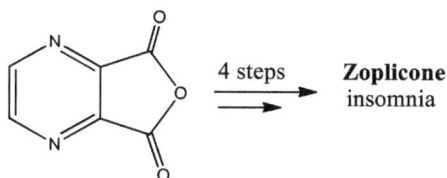 4 steps   **Zoplicone**
insomnia

(**a**) 2-amino-5-chloropyridine heat; (**b**) SOCl$_2$; (**c**) LiAlH$_4$ (1.0 eq.); (**d**)

23)

PhCH$_2$CN   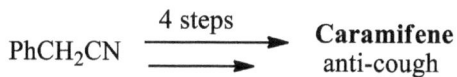   **Caramifene**
anti-cough

(**a**) 2.0 eq. base and 1,2-dibromobutane; (**b**) strong acid or basic medium; (**c**) thionyl chloride; (**d**) HOCH$_2$CH$_2$N(Et)$_2$.

24)

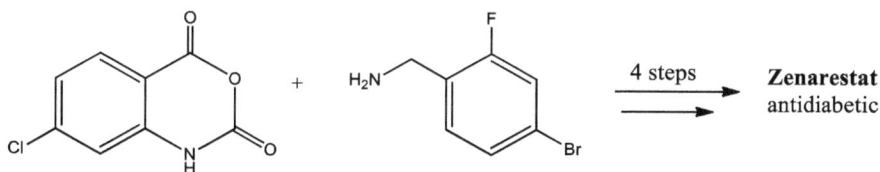 4 steps   **Zenarestat**
antidiabetic

(**a**) heating; (**b**) CDI; (**c**) BrCH$_2$CO$_2$Et; (**d**) NaOH.

25)

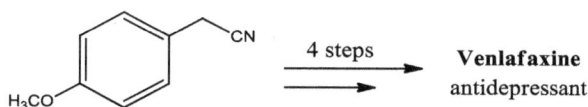 4 steps   **Venlafaxine**
antidepressant

(**a**) *n*-BuLi, THF, -78°C; (**b**) cycloexanone; (**c**) H$_2$-Pd/C; (**d**) 2HCHO, 2HCO$_2$H, H$_2$O, reflux.

The last step in the synthesis of this drug is a methylation reaction. Why was this reaction condition used and not, for example, methyl iodide in basic media?

26)

Gaboxadol
Against insomnia.

(**a**) OHCH₂CH₂OH, TsOH; (**b**) NH₂OH; (**c**) HCl; (**d**) HBr heat, followed by neutralization with Et₃N.

27)

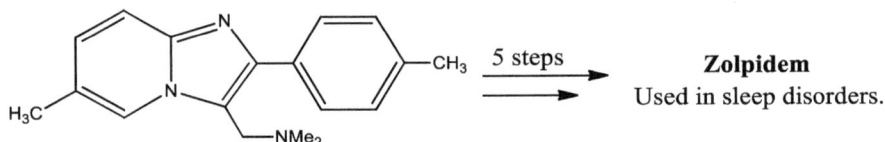

Zolpidem
Used in sleep disorders.

(**a**) MeI; (**b**) KCN; (**c**) HCl reflux; (**d**) SOCl₂; (**e**) dimethylamine.

28)

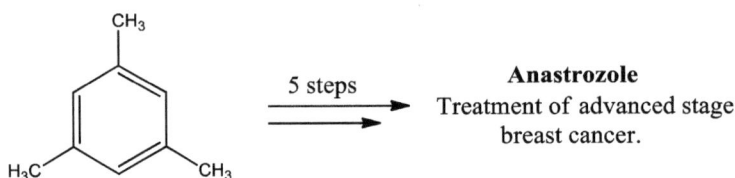

Anastrozole
Treatment of advanced stage breast cancer.

(**a**) 2 eq. NBS, (PhCO)₂O₂ (radical initiator), CCl₄, reflux; (**b**) KCN, DCM/H₂O, reflux; (**c**) NaH, MeI, DMF; (**d**) 1,0 eq. NBS, (PhCO)₂O₂ (radical initiator), CCl₄, reflux; (**e**) 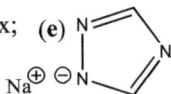, DMF.

29)

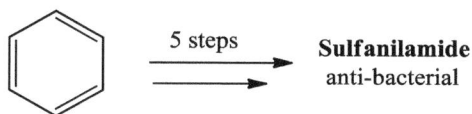

(**a**) nitration; (**b**) reduction; (**c**) sulphonation with fuming sulfuric acid; (**d**) PCl$_5$; (**e**) NH$_3$ or NH$_4$OH.

30)

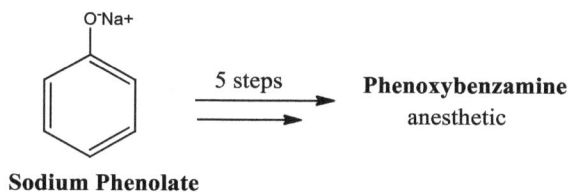

Sodium Phenolate

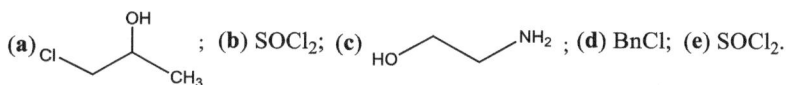

31)

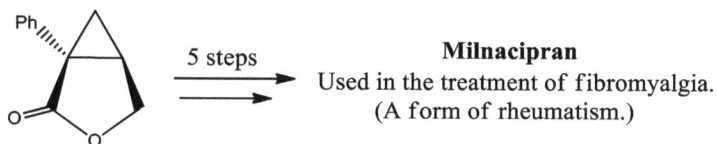

(**a**) potassium phthalimide; (**b**) SOCl$_2$; (**c**) Et$_2$NH; (**d**) NH$_2$NH$_2$; (**e**) HCl.

32)

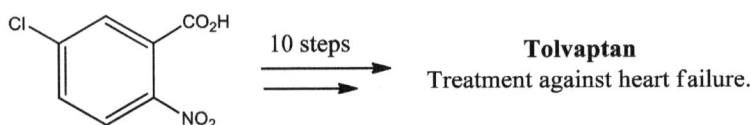

Tolvaptan

Treatment against heart failure.

(**a**) $(CH_3)_2SO_4$; (**b**) $SnCl_2$; (**c**) TsCl; (**d**) $ClCH_2CH_2CH_2CO_2Et$; (**e**) *t*-BuOK; (**f**) HCl (-CO_2);

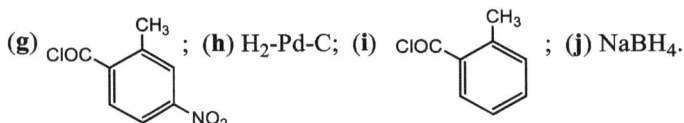

(**g**) ... ; (**h**) H_2-Pd-C; (**i**) ... ; (**j**) $NaBH_4$.

33)

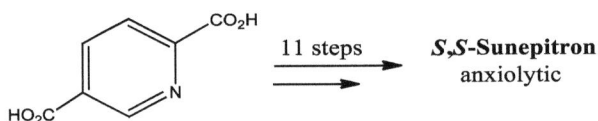

11 steps

S,S-Sunepitron

anxiolytic

(a) 2.0 eq. $SOCl_2$, MeOH (solvent); (b) H_2 and pressure; (c) base followed by $ClCH_2CN$; (d) H_2-Ni-Raney; (e) 2.0 eq. $LiAlH_4$; (f) base followed by 2-chloropyrimidine; (g) mesyl chloride and base; (h) NaN_3; (i) H_2-Pd-C; (j) resolution of diastereoisomers; (k) succinic anhydride.

Answers

1)

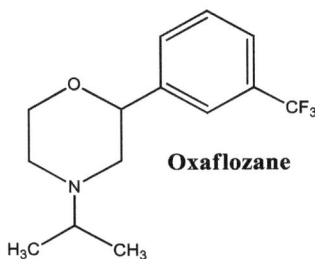

Oxaflozane

2) Dehydration of benzyl alcohol due to the formation of a more stable carbocation.

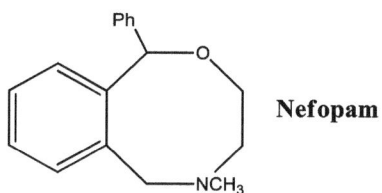

Nefopam

3)

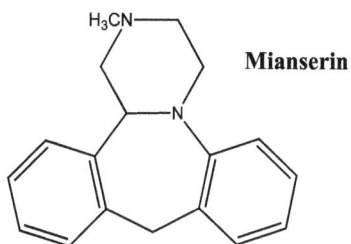

Mianserin

4) Chlorine in aromatic systems is a better leaving group.

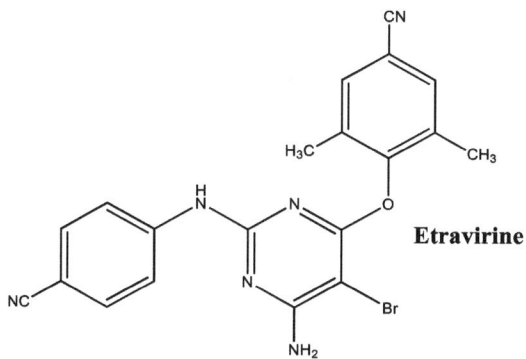

Etravirine

5)

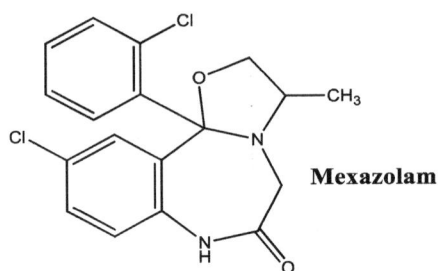

Mexazolam

6)

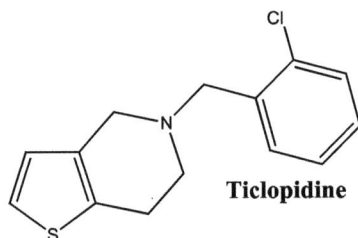

Ticlopidine

7)

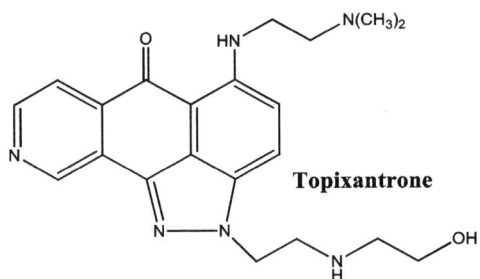

Topixantrone

8)

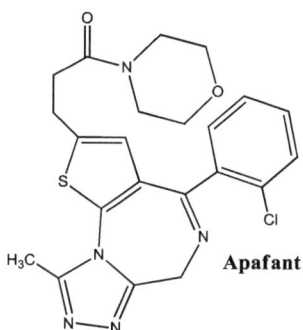

Apafant

9)

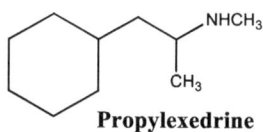

Propylexedrine

10) Mannich reaction. Piperidine also serves as the basis for the reaction.

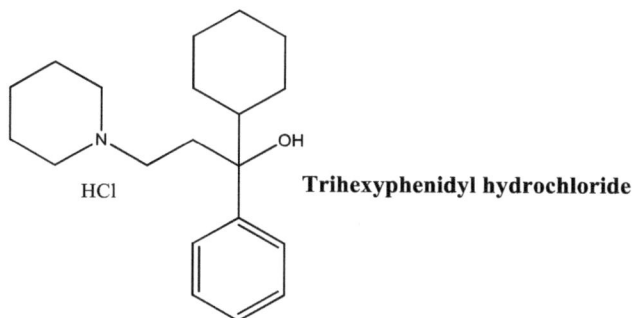

HCl   **Trihexyphenidyl hydrochloride**

11) Mannich reaction. *N, N*-dimethylamine also serves as the base in the reaction and is also responsible for the elimination reaction after β-amination formation.

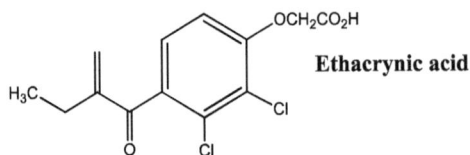

Ethacrynic acid

12) Sulfonation reaction, using chlorosulfonic acid, is oriented by the *ortho-*effects of the chlorine atoms. The nucleophilic aromatic substitution is preferably given in the *ortho* position to the carboxylic acid, as it is an electron-withdrawing group.

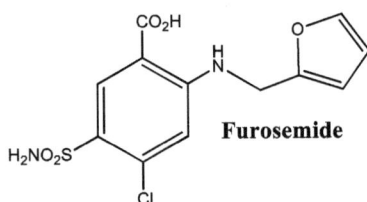

Furosemide

13)

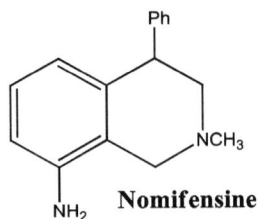

Nomifensine

14)

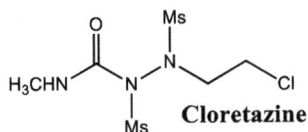

Cloretazine

15) The use of the first step is to form the *N*-oxide by activating the nitration in the para position. The last step is the reduction simultaneously of the nitro and the *N*-oxide groups.

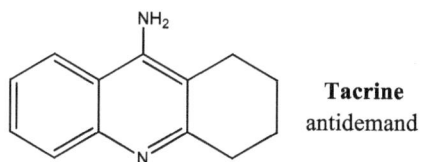

Tacrine
antidemand

16)

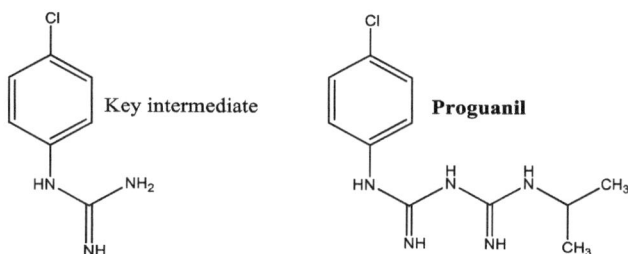

Key intermediate

Proguanil

17)

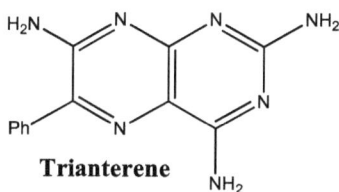

Trianterene

18)

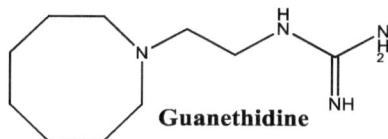

Guanethidine

19) (**a**) intramolecular Friedel-Crafts reaction; (**b**) nucleophilic addition to carbonyl group; (**c**) breaking the OH bond by hydrogen.

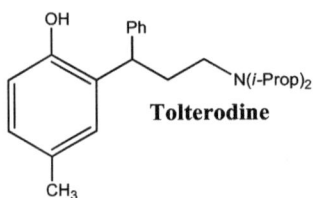

Tolterodine

20) (a) acetylation of the amine; **(b)** introduction of the sulfonyl chloride in the *para* position; **(c)** formation of the sulfonamide; **(d)** deprotection of the amine.

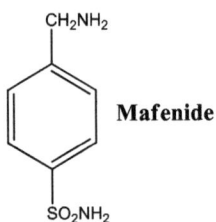

Mafenide

21) (a) bromination; **(b)** elimination reaction (formation of the double bond); **(c)** Grignard reaction; **(d)** dehydration of the alcohol function with the formation of a double bond.

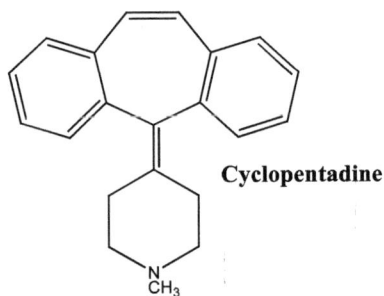

Cyclopentadine

22)

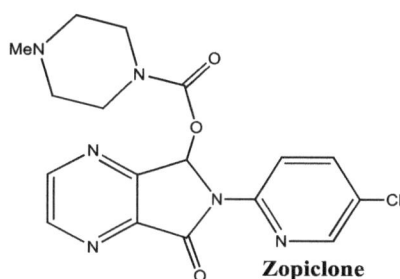
Zopiclone

23) (a) formation of the 5-membered ring; **(b)** hydrolysis of the nitrile to the carboxylic acid function; **(c)** formation of the acid chloride; **(d)** ester formation.

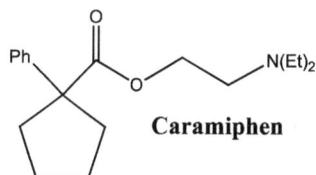
Caramiphen

24) (a) coupling reaction of the two intermediates (formation of the amide concomitant with the decarboxylation); (b) introducing a carbon atom to form the six-membered ring; (c) *N*-alkylation reaction; (d) saponification reaction.

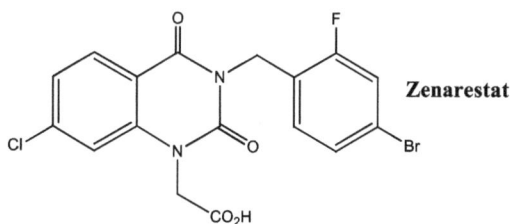
Zenarestat

25) (a) formation of a negative charge α to the cyano group; **(b)** carbonyl addition reaction; **(c)** reduction of nitrile; **(d)** *N,N*-dimethylation. The reaction using methyl iodide will not be selective, and the methylation of the hydroxyl group may also occur. Already in the case of the condition used, it is selective for the amine function, because it will form an imine, which will then be reduced by the formic acid ($H_2CO_2 \rightarrow H_2 + CO_2$).

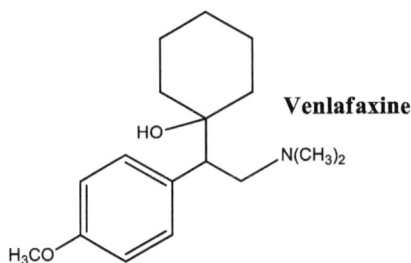

Venlafaxine

26) (a) protection of carbonyl; **(b)** formation of *N*-hydroxyamide; **(c)** deprotection reaction followed by the formation of the five-membered ring and elimination reaction (intramolecular aldol condensation type reaction); **(d)** *N*-deacetylation (a more drastic acidic condition is required).

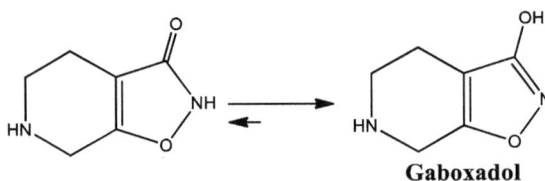

Gaboxadol

27) (a) formation of the quaternary amine salt; **(b)** introducing the nitrile function due to the formation of the quaternary amine salt which functions as the leaving group; **(c)** transforming nitrile into carboxylic acid; **(d)** acid chloride; **(e)** amide formation.

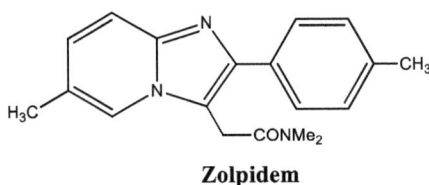

Zolpidem

28) The reader should remember to observe the elements of symmetry in the substance. **(a)** dibromination at benzyl carbons; **(b)** reaction of type SN2; **(c)** methylation at α-nitrile position; **(d)** bromination on benzylic carbon; **(e)** reaction of type SN2.

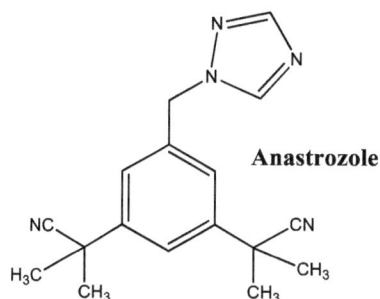

Anastrozole

29)

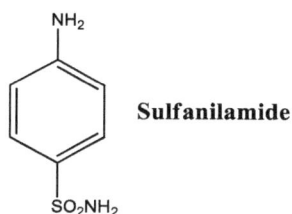

Sulfanilamide

30) Thionyl chloride is also capable of converting hydroxyls to chlorides. The mechanism of the reaction is similar when using this reagent in the conversion of carboxylic acids to acid chlorides.

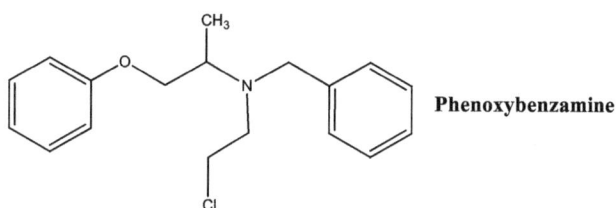

Phenoxybenzamine

31) (a) opening of the lactone with the potassium salt of phthalimide, classical method of obtaining amines known as Gabriel synthesis; **(b)** transforming the carboxylic acid into acid chloride; **(c)** amide formation; **(d)** cleavage of the phthalimide group to obtain the chemical amine function; **(e)** formation of the hydrochloride.

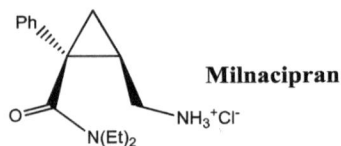

Milnacipran

32) **(a)** esterification; **(b)** reduction of the nitro group; **(c)** tosylation of the amine; **(d)** alkylation reaction of the sulfonamide chemical function; **(e)** formation of a seven-membered ring (abstraction of α-carbonylic hydrogen, followed by its nucleophilic attack on the ester function which is directly attached to the aromatic ring, thus producing a β-ketoester); **(f)** decarboxylation of the ester function concomitantly with cleavage of the tosyl group; **(g)** formation of the chemical amide function; **(h)** reduction of the nitro group; **(i)** formation of the chemical amide function; **(j)** carbonyl reduction.

33) **(a)** esterification (acid chloride + alcohol); **(b)** reduction of the pyridine nucleus; **(c)** *N*-alkylation; **(d)** reducing the chemical nitrile function in the amine followed by the formation of a lactam; **(e)** reducing the ester in alcohol and the lactam in amine; **(f)** aromatic nucleophilic substitution reaction between the formed amine and 2-chloropyrimidine; **(g)** transforming the hydroxyl group into a good leaving group (mesyl group); **(h)** SN2 reaction; **(i)** reduction of the azido group; **(j)** separating the diastereoisomers; **(k)** formation of a succinimide.

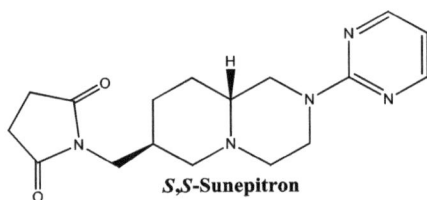

S,S-Sunepitron

EXERCISE 8 - COMPLETING THE GAPS

The exercises below are based on a series of gaps in the synthesis of different drugs. Therefore, it is up to the reader to complete them. This type of exercise is a combination of the themes previously presented. It is necessary to identify the reagents' use, the reaction conditions employed, and the intermediates formed, besides the use of the concept of retrosynthesis.

Example

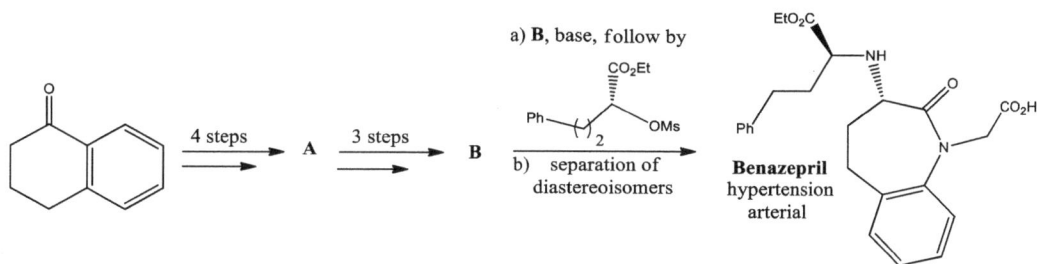

a) **B**, base, follow by

b) separation of diastereoisomers

Benazepril
hypertension
arterial

Answer

As we insisted, it is imperative in any exercise always to perform the retrosynthetic analysis. This methodology is undoubtedly the most efficient and rapid way of making coherent synthetic proposals. The study of the drug benazepril in the example above proposes two starting materials. The critical part is to identify the formation of the seven-membered ring. This step occurs using a rearrangement, usually known as Beckmann's rearrangement, which is the transformation of the oxime functional group to substituted amides. In the cyclic systems, these reactions occur with ring expansion; in this case, six to seven-membered ring. Another detail is the introduction of the amino group in the α-carbonyl position *via* the synthesis of Gabriel. This synthesis converts alkyl halides into amides by using salt of phthalimide and hydrazine.

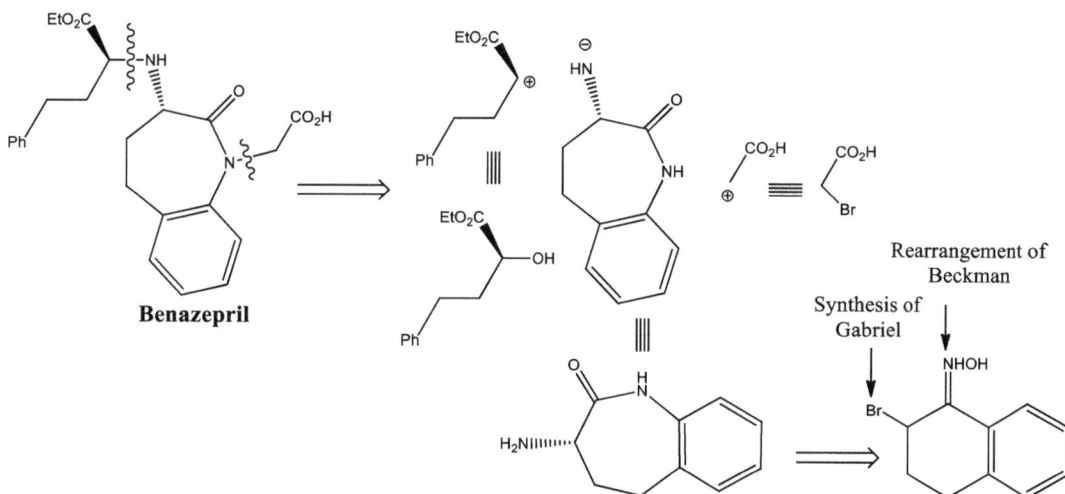

Benazepril

Rearrangement of Beckman

Synthesis of Gabriel

Rearrangement of
Beckman

Ring expansion

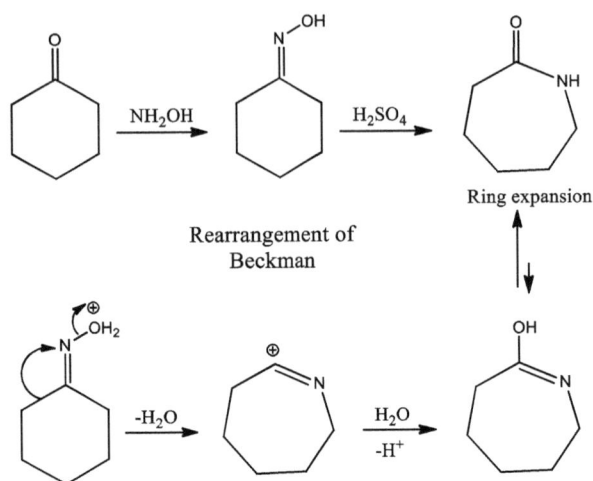

Synthesis of Gabriel

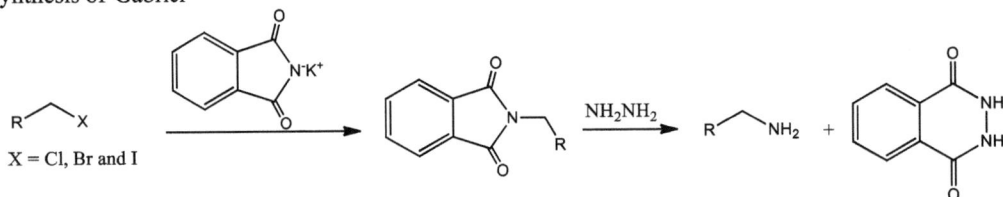

X = Cl, Br and I

(**a**) base; (**b**) Br$_2$; (**c**) NH$_2$OH; (**d**) H$_3$PO$_4$; (**e**) potassium phthalimide; (**f**) BrCH$_2$CO$_2$*t*-Bu; (**g**) ethanolamine or hydrazine.

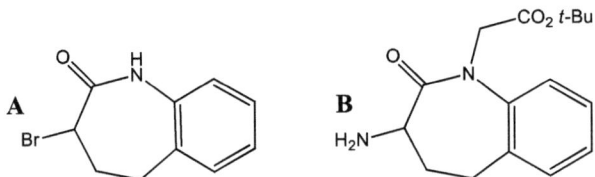

A   **B**

EXERCISES

1)

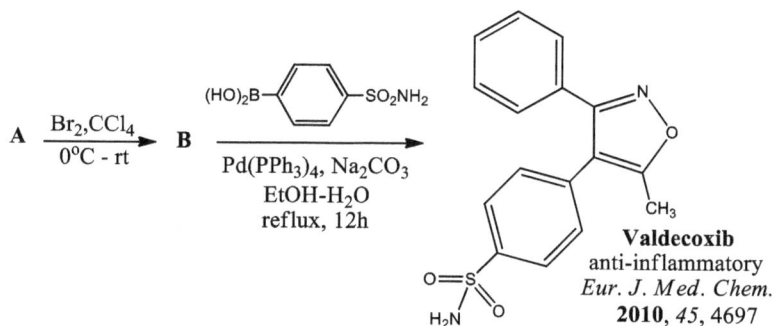

Valdecoxib
anti-inflammatory
Eur. J. Med. Chem.
2010, *45*, 4697

2) Tip: Methylation occurs in sulfur.

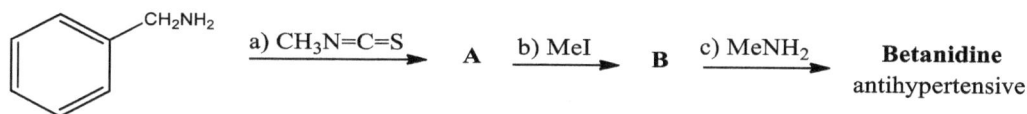

Betanidine
antihypertensive

3)

$(Ph)_2CO$ a) $NaBH_4$ **Diphenylpyraline**
           b) $SOCl_2$ antihistamine

          c) H_3CN —OH

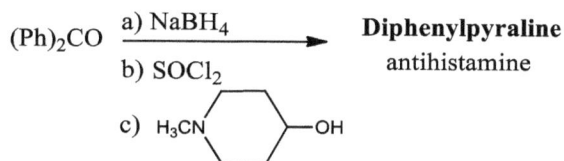

4)

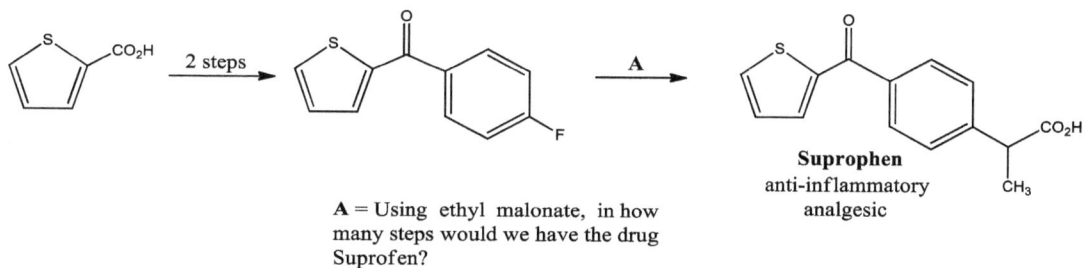

A = Using ethyl malonate, in how many steps would we have the drug Suprofen?

Suprophen
anti-inflammatory
analgesic

5)

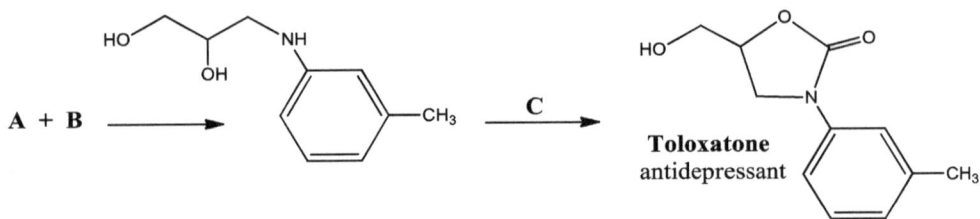

A + B ⟶ [intermediate] —C→ **Toloxatone** antidepressant

6)

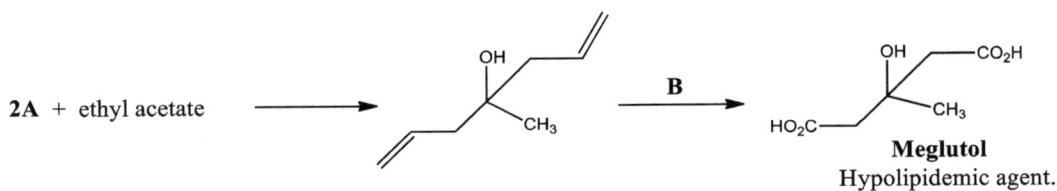

2A + ethyl acetate ⟶ [intermediate] —B→ **Meglutol** Hypolipidemic agent.

7)

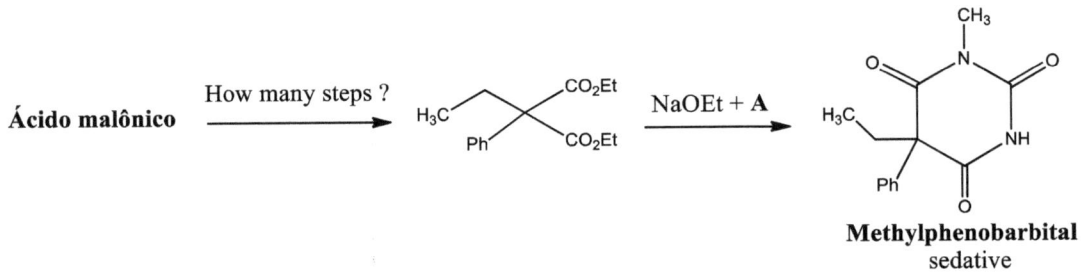

Ácido malônico —How many steps ?→ [intermediate] —NaOEt + A→ **Methylphenobarbital** sedative

8)

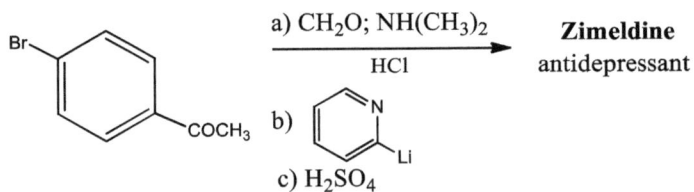

a) CH$_2$O; NH(CH$_3$)$_2$
HCl
b) [2-lithiopyridine]
c) H$_2$SO$_4$

⟶ **Zimeldine** antidepressant

9)

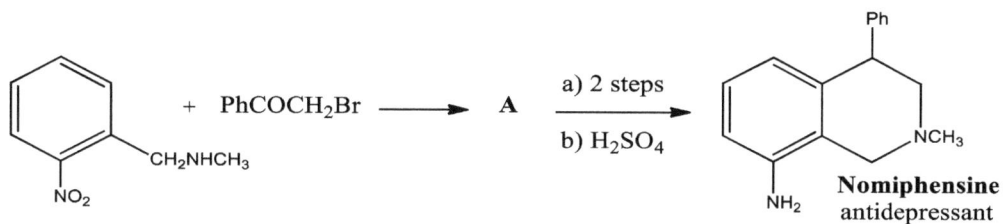

10)

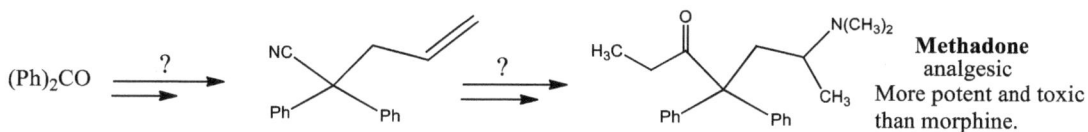

11)

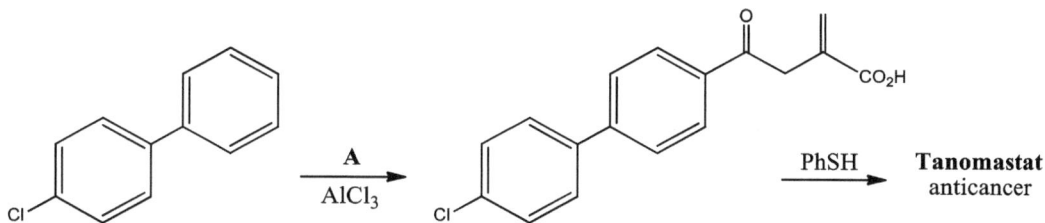

12)

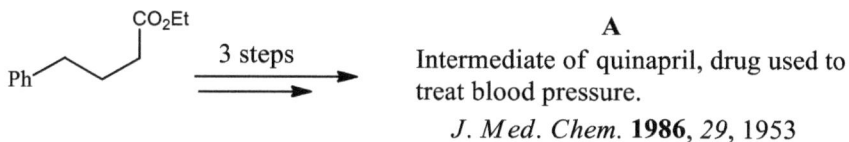

CO_2Et Ph — (3 steps) → **A**
Intermediate of quinapril, drug used to treat blood pressure.

J. Med. Chem. **1986**, *29*, 1953

(**a**) Base, bromine; (**b**), H_2N — CH_3 / CO_2t-Bu Et_3N; (**c**) HCl $_{(gas)}$, THF, DCM

13)

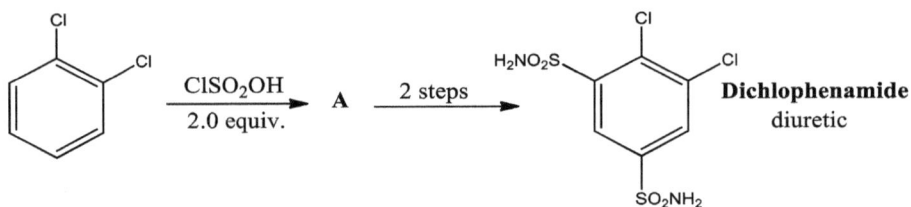

14)

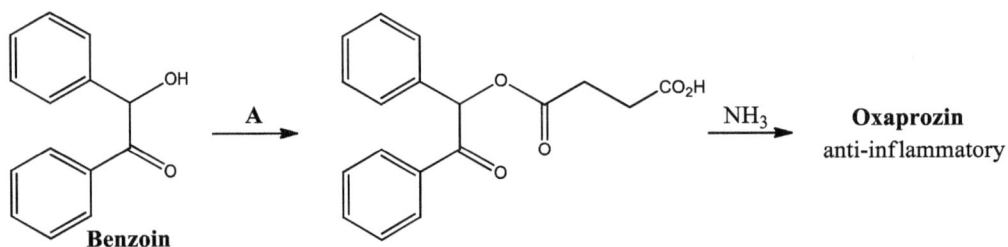

15)

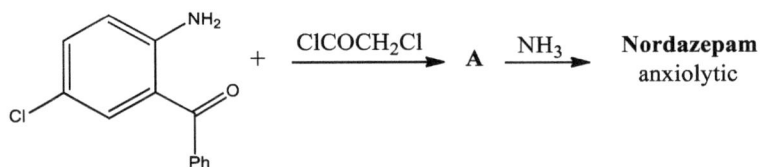

16)

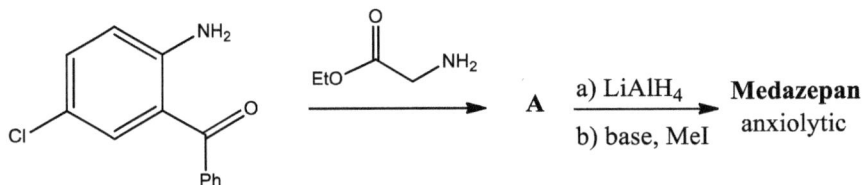

17)

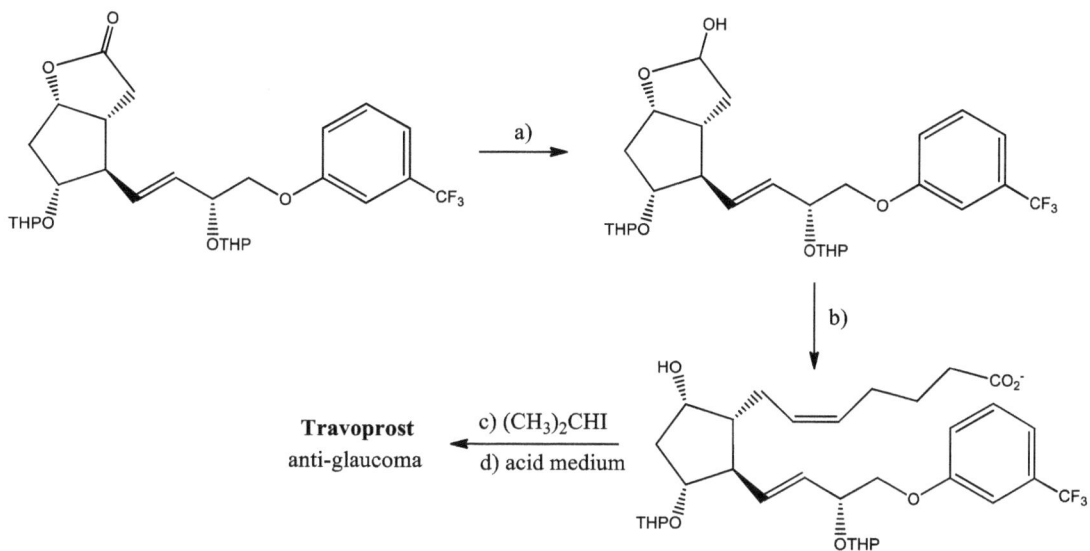

Travoprost
anti-glaucoma

c) $(CH_3)_2CHI$
d) acid medium

18)

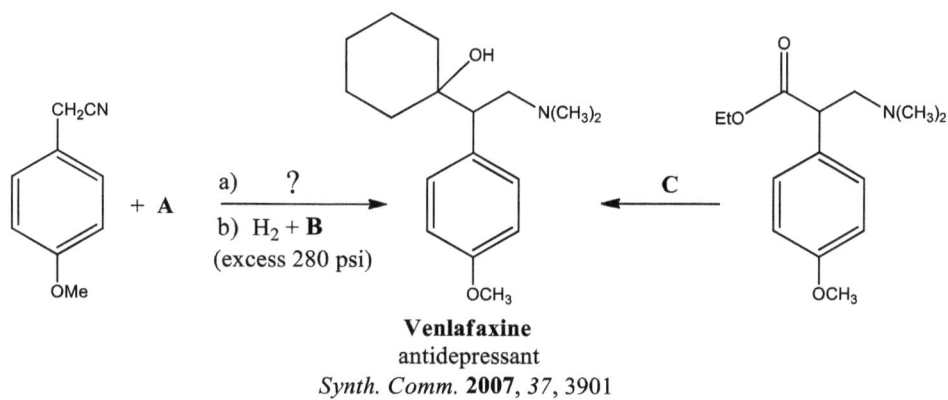

a)   ?
b) H_2 + **B**
(excess 280 psi)

C

Venlafaxine
antidepressant
Synth. Comm. **2007**, *37*, 3901

19)

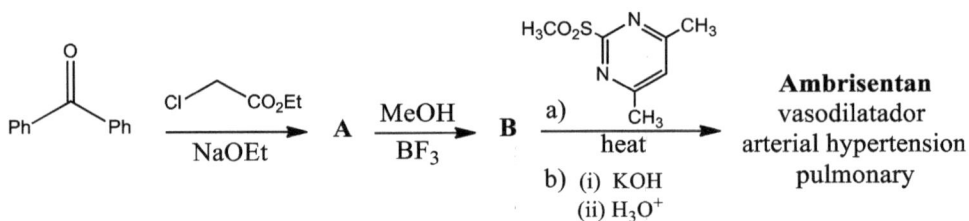

Ambrisentan
vasodilatador
arterial hypertension
pulmonary

20)

Thiamazole
hyperthyroidism

21)

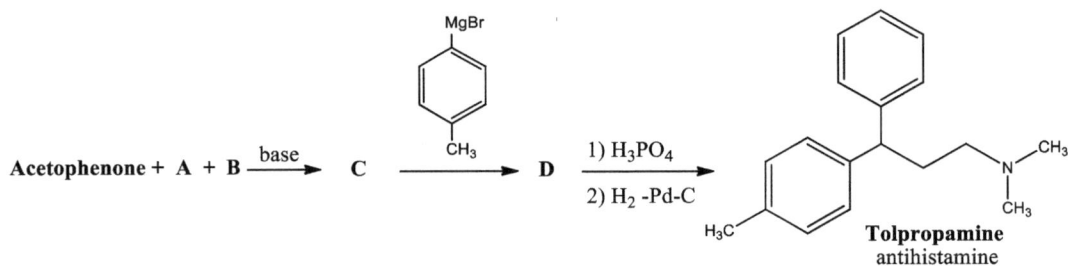

Tolpropamine
antihistamine

22)

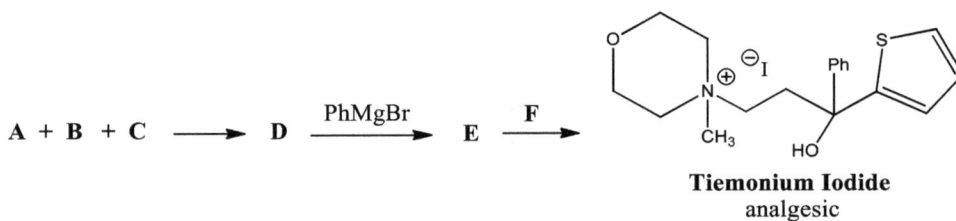

Tiemonium Iodide
analgesic

23)

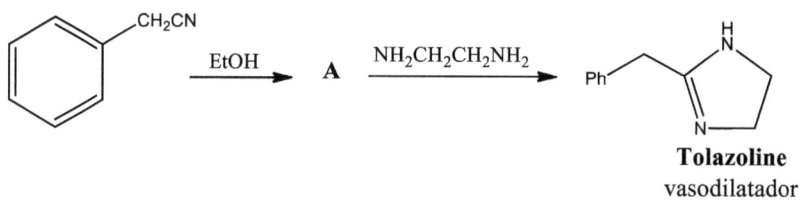

Tolazoline
vasodilatador

24)

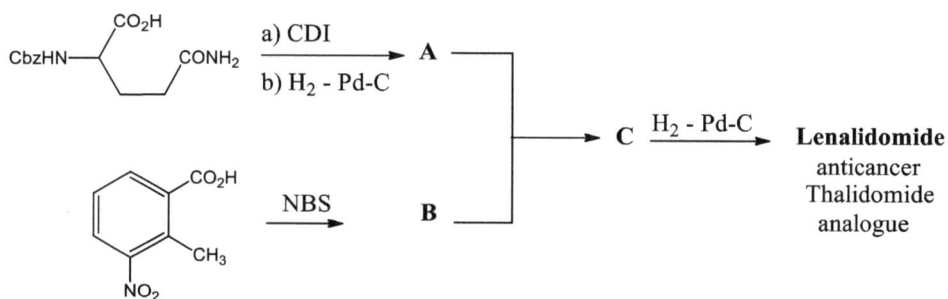

Lenalidomide
anticancer
Thalidomide
analogue

25)

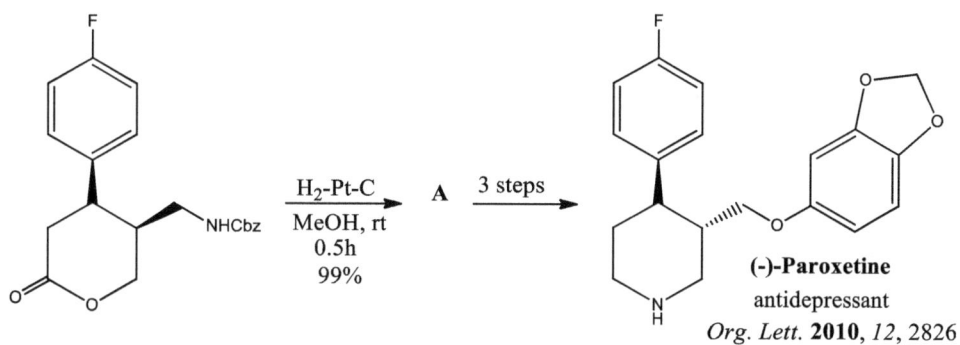

(-)-Paroxetine
antidepressant
Org. Lett. **2010**, *12*, 2826

26)

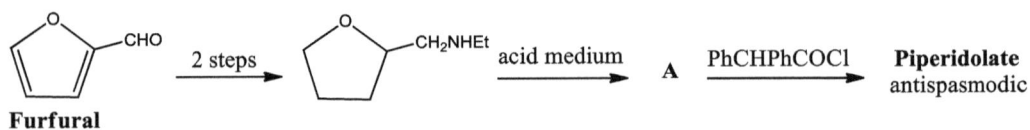

Furfural —CHO → (2 steps) → CH₂NHEt → (acid medium) → **A** → (PhCHPhCOCl) → **Piperidolate** antispasmodic

27)

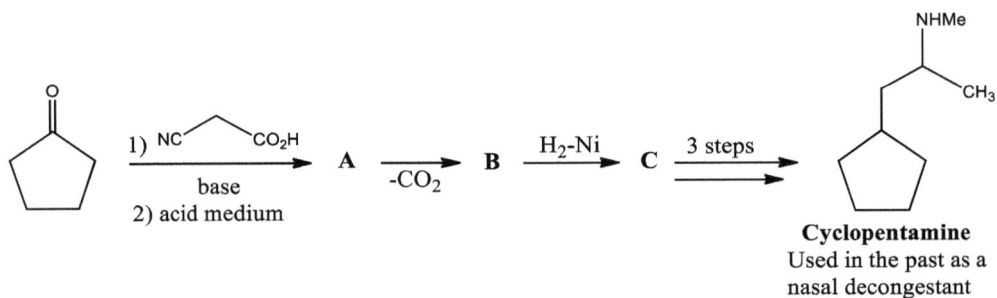

1) NC—CO₂H / base
2) acid medium → **A** → (-CO₂) → **B** → (H₂-Ni) → **C** → (3 steps) → **Cyclopentamine** Used in the past as a nasal decongestant

28) In step (**h**), is the use of the catalyst important for the success of the reaction? Justify your answer.

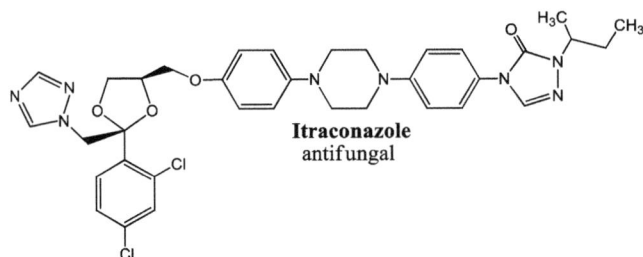

Itraconazole antifungal

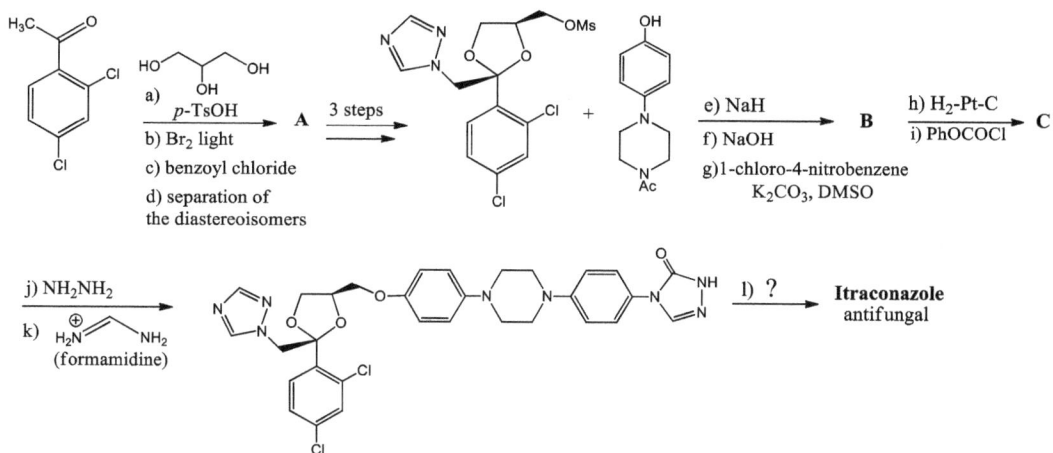

29)

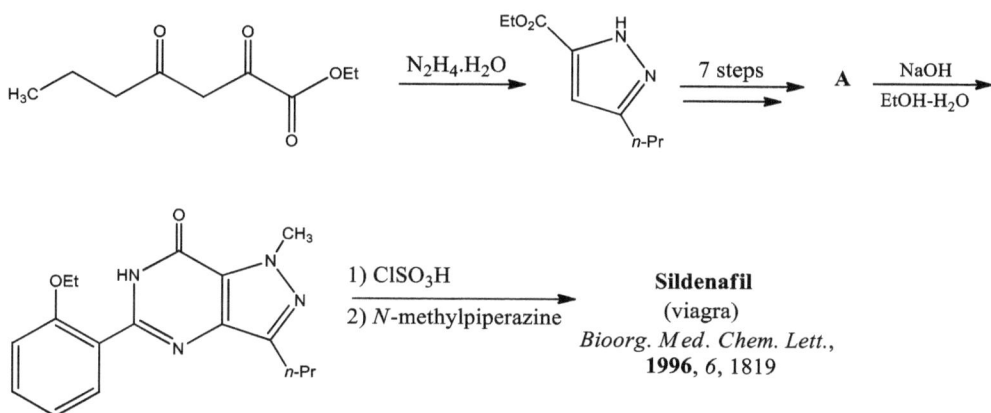

1) ClSO$_3$H
2) *N*-methylpiperazine

Sildenafil
(viagra)
Bioorg. Med. Chem. Lett.,
1996, *6*, 1819

30)

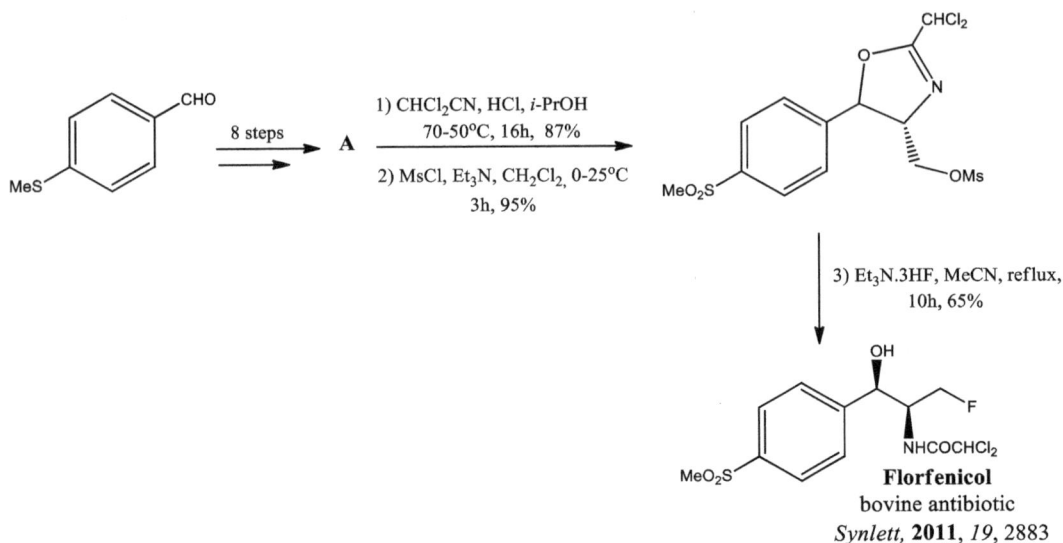

Florfenicol
bovine antibiotic
Synlett, **2011,** *19,* 2883

(**a**) MeO$_2$CCH$_2$P(=O)(OEt)$_2$, NaH, THF, 0–25°C, 2.5 h, 92%; (**b**) LiAlH$_4$, AlCl$_3$, Et$_2$O, 0–25°C, 2 h, 90%; (**c**) L-DIPT, Ti(Oi-Pr)$_4$, t-BuOOH, molecular sieve 4Å, CH$_2$Cl$_2$, 40°C, 70%, 75% ee; (**d**) BnBr, NaH, DMF, –20°C, 5 h, 91%; (**e**) BnOH, BF$_3$·OEt$_2$, rt, 5 h, 70%; (**f**) MsCl, Et$_3$N, CH$_2$Cl$_2$, 0–25°C, 3 h; (**g**) NaN$_3$, DMF, 100°C, 10 h, 75% in two steps; (**h**) 10% Pd/C, H$_2$, MeOH, rt 48 h, 85%.

31)

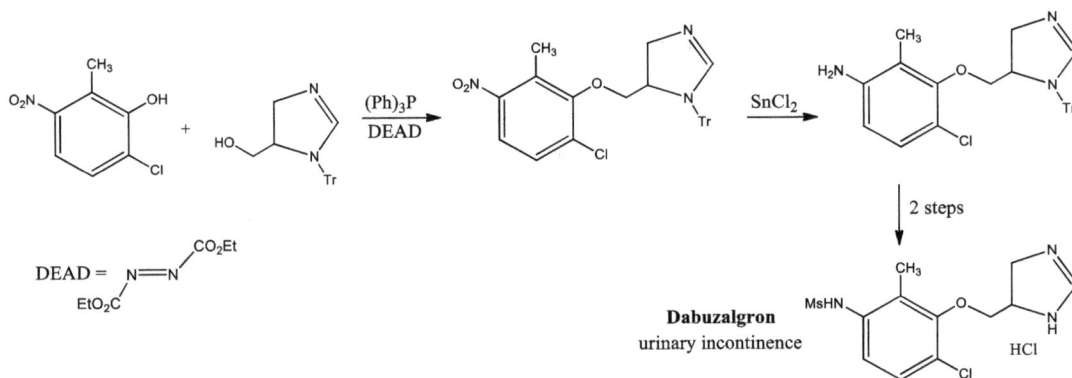

Dabuzalgron
urinary incontinence

1) Why was the known Mitsunobu reaction used instead of mesyl chloride (CH$_3$SO$_2$Cl)?

2) What is the mechanism of this reaction?

3) Why was the reduction reaction of the nitro functional group not carried out using classical conditions such as hydrogen with palladium as a catalyst or Fe-HCl?

Answers

1)

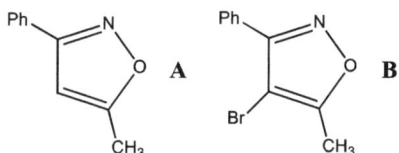

2)

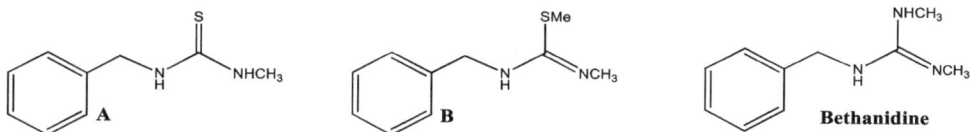

3)

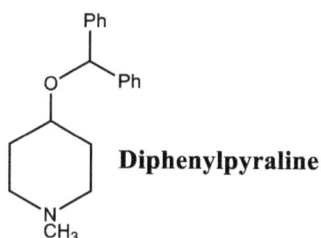

Diphenylpyraline

(a) reduction of the ketone to alcohol (NaBH$_4$); **(b)** transforming the formed alcohol into a good leaving group (mesylate, chlorine or bromine); **(c)** coupling with 1-methyl-4-piperidinol.

4) (a) thionyl chloride; **(b)** aluminum (III) chloride and fluorobenzene; **A** = **(1)**

ethyl malonate, base (sodium ethoxide) and methyl iodide (**2**) base, followed by aromatic nucleophilic substitution, saponification and decarboxylation.

5) A = glycidol; **B** = 3-methylaniline; **C** = dimethyl carbonate $CO(OMe)_2$ and sodium methanolate.

6) A = allyl magnesium bromide; **B** = ozone and hydrogen peroxide.

7) 3 steps = (**a**) esterification; (**b**) base followed by ethyl iodide; (**c**) base followed by fluorobenzene.

A = *N*-methylurea

8)

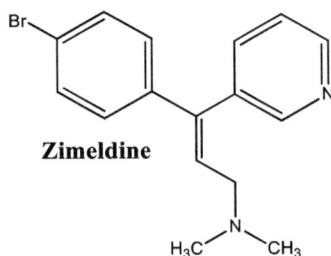

Zimeldine

9)

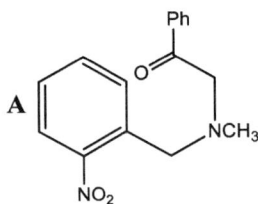

(**a**) H_2 - Níquel Raney; (**b**) sodium borohydride.

10) A = (**a**) reduction of the ketone to alcohol ($NaBH_4$); (**b**) transforming the formed alcohol into a good leaving group (sulfonate or halide); (**c**) nucleophilic substitution (KCN); (**d**) base; (**e**) allyl bromide.

B = (**a**) HBr; (**b**) diethylamine; (**c**) EtMgBr; (**d**) slightly acidic medium (steps **c** and **d** are for the conversion of the nitrile functional group to ketone).

11)

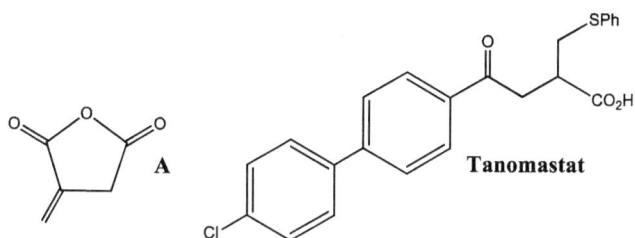

A   **Tanomastat**

12)

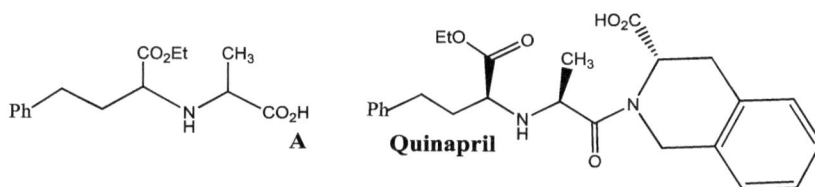

A   **Quinapril**

Selective deprotection of the *t*-Bu group in an acid medium, due to the formation of a more stable tertiary carbocation.

13)

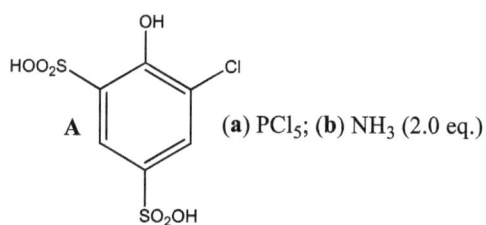

A   **(a)** PCl$_5$; **(b)** NH$_3$ (2.0 eq.)

14)

A = succinic anhydride

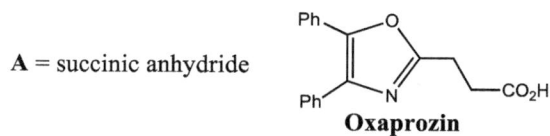

Oxaprozin

15)

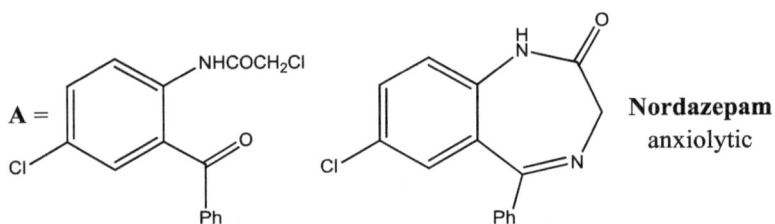

A =

Nordazepam
anxiolytic

16)

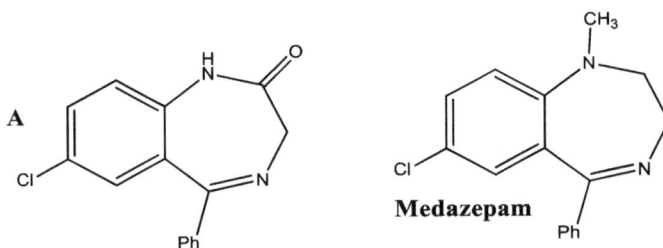

A

CH$_3$

Medazepam

17)

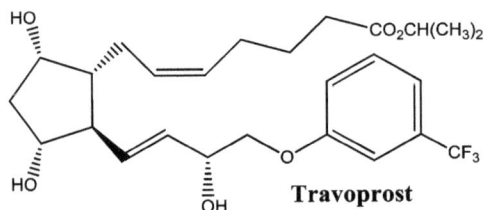

Travoprost

(a) DIBAL-H; **(b)** (Ph)$_3$P$^+$(CH$_2$)$_3$CO$_2^-$.

18) A = cycloexanone; **B** = HCHO (2.0 eq.); **C** = BrMg(CH$_2$)$_5$MgBr.

(a) base; **(b)** H$_2$, 280 psi, aldehyde, MeOH, 100°C, 30% (simultaneous reduction of nitrile function and reductive amination).

19)

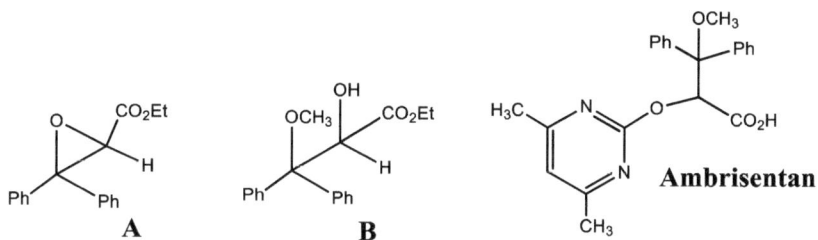

A   **B**   **Ambrisentan**

20)

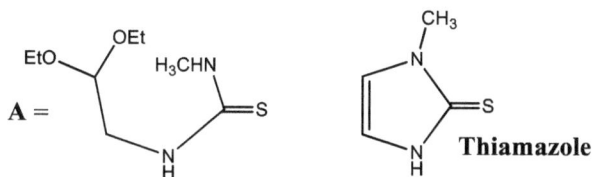

A =   **Thiamazole**

21)

A = formaldehyde; **B** = dimethylamine; **C** =   ; **D** =
(Mannich reaction)

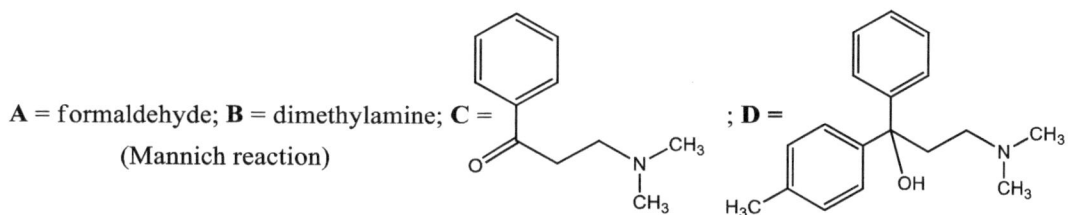

22)

A = morpholine; **B** = formaldehyde; **C** = 2-acetylthiophene; **D** =   ; **E** =

F = methyl iodide.

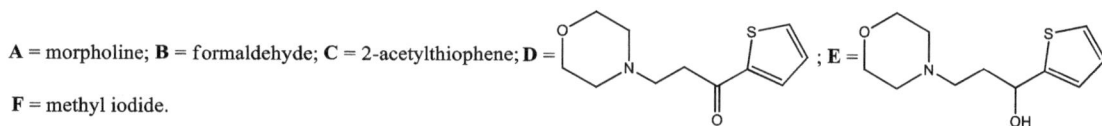

23)

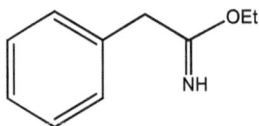

24)

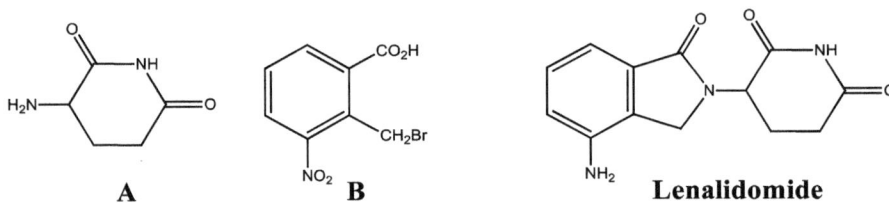

    A   **B**   **Lenalidomide**

25) (a) MsCl, DMAP, Et$_3$N, CH$_2$Cl$_2$, rt, 1d; **(b)** NaH, DMF, sesamol, reflux, 0.5h, 56% in two steps; **(c)** LiAlH$_4$, Et$_2$O, rt, 1h, 90%.

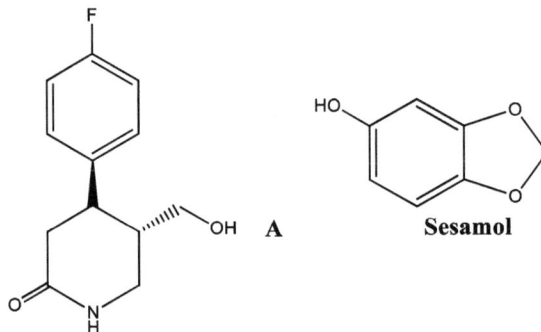

    A   **Sesamol**

26) Reaction condition of the first step: NH$_2$Et; H$_2$-Pd-C (pressure) (reductive amination and homogenization of the heteroaromatic system).

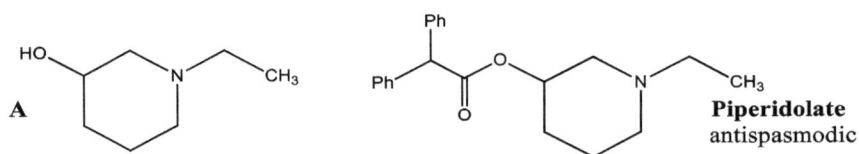

    A   **Piperidolate**
                                   antispasmodic

27) (a) MeMgBr, followed by mildly acid medium; **(b)** methylamine; (c) H$_2$-Pt.

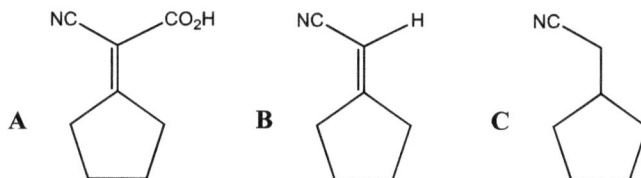

To obtain product **C**, it is necessary to use nickel as a catalyst to not hydrogenate the chemical nitrile function.

28)

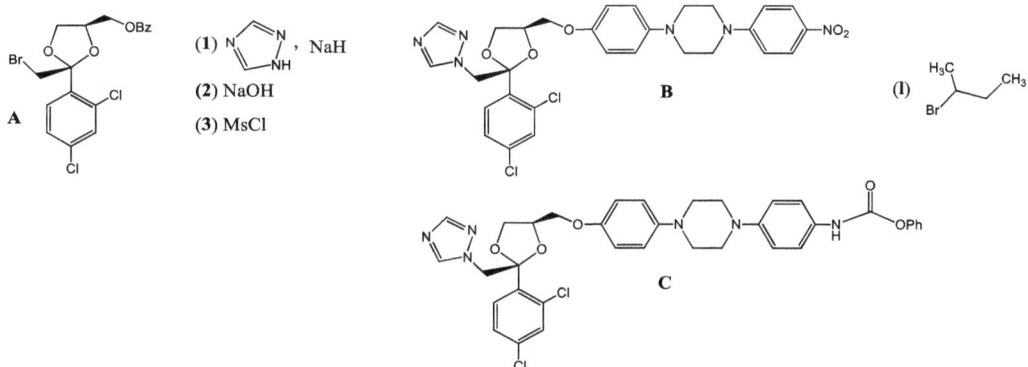

The use of the catalyst is crucial, for example, if palladium is used instead of platinum, halogenolysis of the chlorine atoms present in the aromatic ring would occur.

29) (a) Base followed by Me$_2$SO$_4$ (in the article only dimethyl sulphate was used without the presence of the base); **(b)** NaOH-H$_2$O followed by slightly acidic medium (saponification reaction); **(c)** HNO$_3$-H$_2$SO$_4$; **(d)** SOCl$_2$; **(e)** NH$_4$OH; **(f)** SnCl$_2$ (reduction reaction).

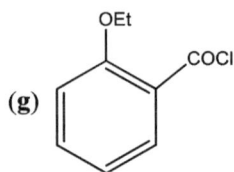

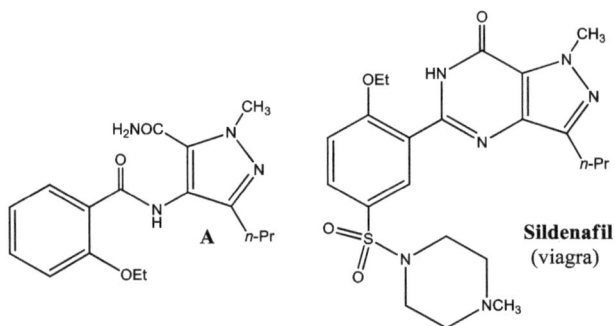

(g)

30)

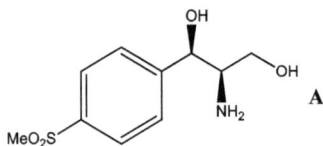

It is noteworthy that in step (**b**), the sulfone will also be reduced and in (**c**), the sulfur will be oxidized back to the sulfone.

31) 1) This is due to the mesyl chloride being able to react with the phenolic hydroxyl.

2) Mechanism of Reaction

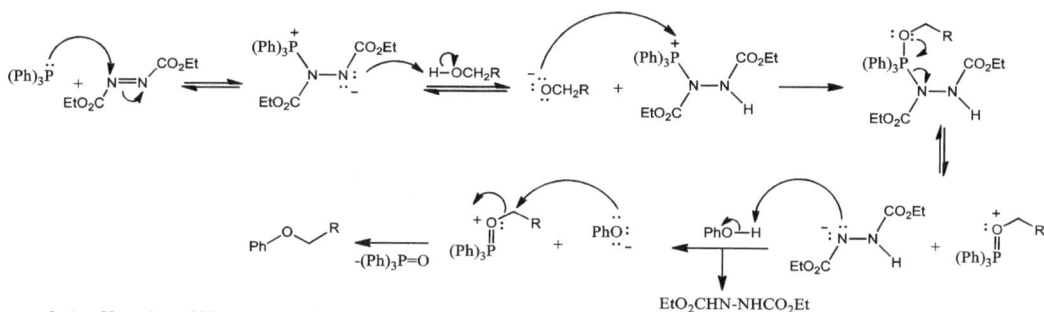

J. Am. Chem. Soc. **1988**, *110*, 6487–6491

It is important to mention that we used the Mitsunobu reaction when one wants to reverse the configuration of a specific stereogenic center (Scheme **1**).

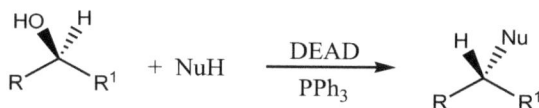

Scheme (1). Mitsunobu reaction.

3) In the reduction process, using H_2-Pd-C, halogenolysis and reduction of the C=N bond of the ring would occur. The use of Fe-HCl would lead to the cleavage of the trityl group.

CONCLUSION OF EXERCISES 7 AND 8 (IDENTIFYING DRUG STRUCTURE AND COMPLETING THE GAPS)

The two exercises in this chapter aim to solidify the concepts already started by exercises **5** and **6**, such as the identification of reactions, reagents, intermediates, and chemical transformations. Therefore, exercise **7** has the objective of identifying the drugs' structure, having information of the starting materials and reagents used. Exercise **8** presents a series of gaps in the synthesis of different drugs, and it is up to the reader to complete them. It is worth mentioning that it is highly recommended that the reader continues using the concept of retrosynthesis in the resolution of the exercises.

REFERENCES

[1]   Kleemann, A.; Engel, J.; Kutscher, B.; Reichert, D. *Pharmaceutical Substances Syntheses Patents Applications,* 2nd; Thieme Stuttgart: New York, **2001**.

[2]   Lednicer, D.; co-workers, *The Organic Chemistry of Drug Synthesis*; John Wiley & Sons, Inc.: New Jersey, **2007**, pp. 1-7.

[3]   Li, J.J.; Johnson, D.S.; Sliskovic, D.R.; Roth, B.D. *Contemporary Drug Synthesis,* 1st ed; John Wiley & Sons, Inc.: New Jersey, **2004**.
      [http://dx.doi.org/10.1002/0471686743]

[4]   Vardanyan, R.; Hruby, V. *Synthesis of Essential Drugs*; Elsevier: Netherlands, **2006**.

[5]   www.drugsyn.org

CHAPTER 4

Heteroaromatic Substances

Abstract: Heteroaromatics are substances having one or more atoms in the aromatic ring other than the carbon atom. This class of compounds is fundamental in developing new drugs, being present in a large number of drugs and bioactive compounds. Some of these cores are listed below, and it is worth noting that due to their immense structural variety, we mentioned only the most common ones. The exercises of this item we based on the formation of different heteroaromatic substances, and it is up to the reader to identify the respective nuclei.

Keywords: Bioactive compounds, Chemical transformations, Drugs, Exercises, Heteroaromatic compounds, Medicinal chemistry, Molecules, Reaction conditions, Reagents, Synthesis, Retrosynthetic analysis, Substances, Synthons.

FIVE-MEMBERED RINGS

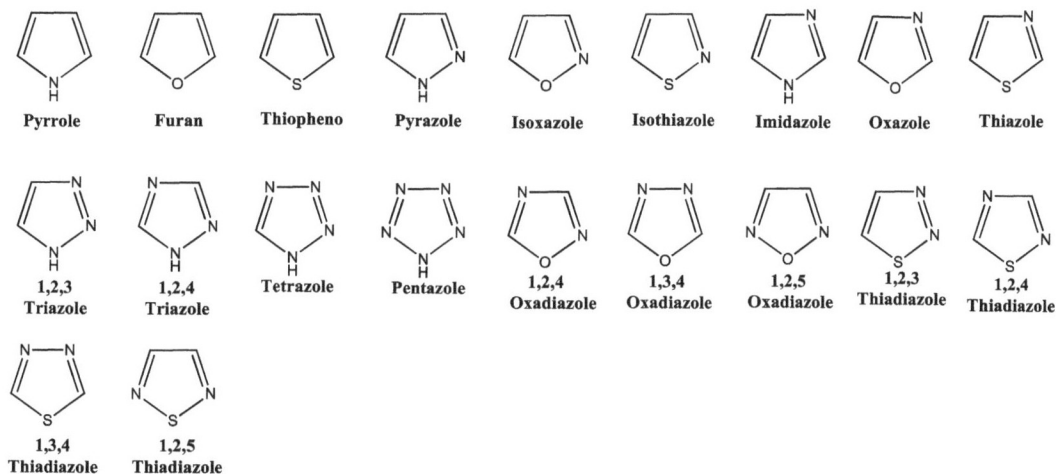

| Pyrrole | Furan | Thiopheno | Pyrazole | Isoxazole | Isothiazole | Imidazole | Oxazole | Thiazole |

| 1,2,3 Triazole | 1,2,4 Triazole | Tetrazole | Pentazole | 1,2,4 Oxadiazole | 1,3,4 Oxadiazole | 1,2,5 Oxadiazole | 1,2,3 Thiadiazole | 1,2,4 Thiadiazole |

| 1,3,4 Thiadiazole | 1,2,5 Thiadiazole |

SIX-MEMBERED RINGS

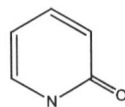 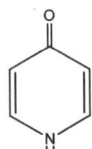 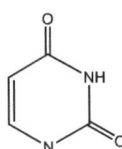 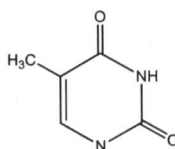 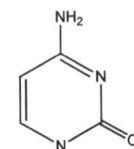 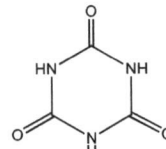

Pyridine   **Pyrimidine**   **Pyrazine**   **Pyridazine**   **1,2,3 Triazine**   **1,2,4 Triazine**   **1,2,5 Triazine**   **1,2,3,5 Tetrazine**

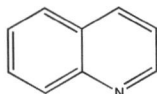 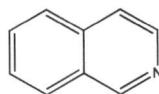 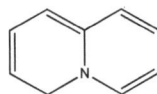 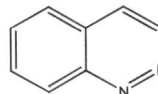 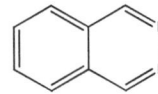

1,2,4,5 Tetrazine   **2-Pyridone**   **4-Pyridone**   **Uracil**   **Thymine**   **Cytosine**   **Cyanuric acid**

FUSED SYSTEMS

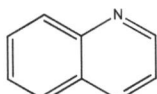 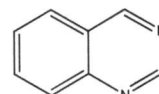 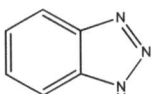 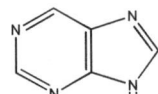 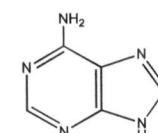

Quinoline   **Isoquinoline**   **Quinolizidine**   **Cinnoline**   **Phthalazine**

Quinoxaline   **Quinazoline**   **Benzotriazole**   **Purine**   **Adenine**

X = NH - Indole
O - Benzofuran
S - Benzothiophene

X = NH - Isoindole
O - Isobenzofuran
S - Benzo[c]thiopheno

X = NH - 1*H*-benzimidazole
O - Benzoxazole
S - Benzothiazole

X = NH - 1*H*-Indazole
O - 1,2-Benzisoxazole
S - 1,2-Benzisothiazole

X = O - 2,1-Benzisoxazole
S - 2,1-Benzisothiazole

Indolizidine

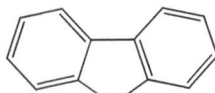 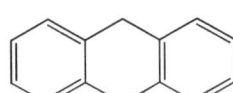 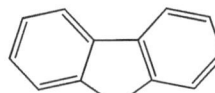

Carbazole   **Acridine**   **Dibenzofuran**

Example

The exercises below indicate the chemical structures of the nuclei formed.

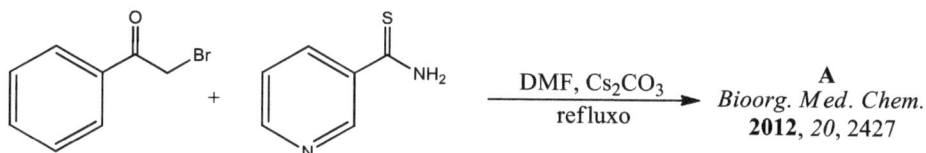

$$\xrightarrow[\text{refluxo}]{\text{DMF, Cs}_2\text{CO}_3}$$

A
Bioorg. Med. Chem.
2012, *20*, 2427

Answer

An important point concerning heteroaromatics is that despite their diversity, the mechanisms of their formation are somewhat similar, so it is fundamental to understand these mechanisms since the reader can easily propose the formation of a specific heteroaromatic nucleus. For example, in general, the structure of these nuclei is almost always related to an intramolecular cyclization reaction followed by the aromatization of the water loss system or an excellent leaving group. The formation of the thiazole nucleus **A** could have occurred according to the mechanism below.

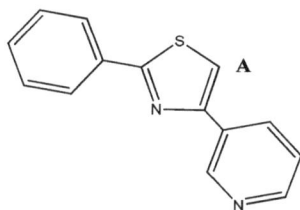

MECHANISM OF REACTION

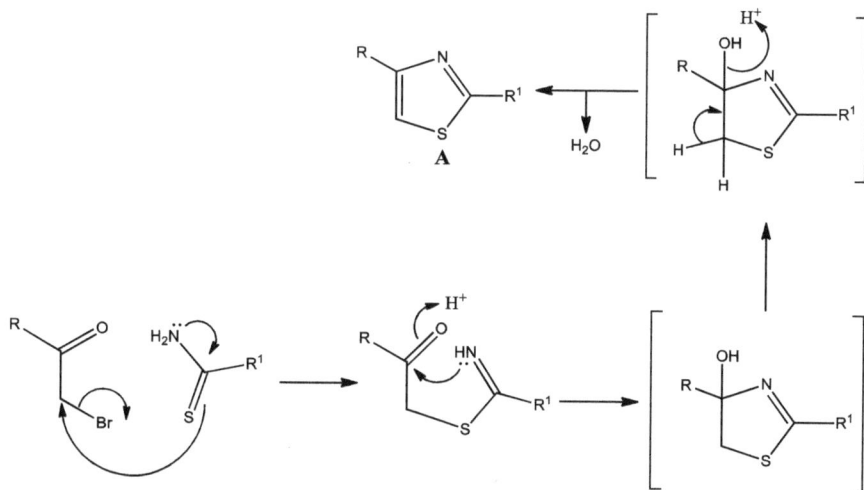

Quim. Nova **2005**, *28*, 77

EXERCISES

The exercises below indicate the chemical structures of the nuclei formed.

1)

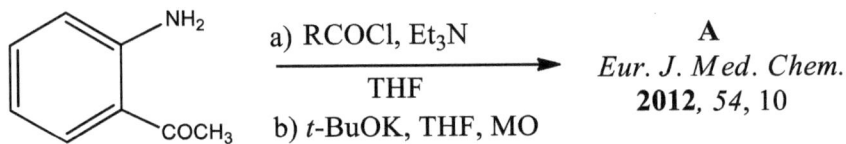

2)

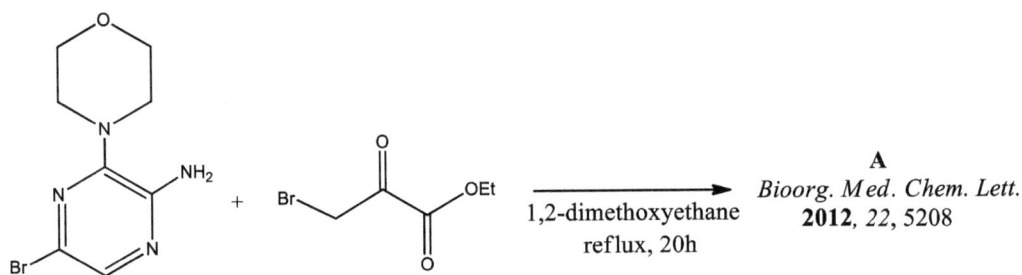

3)

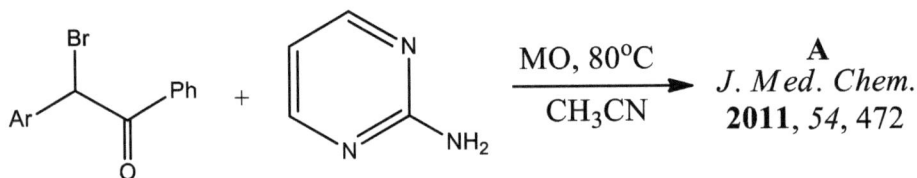

4)

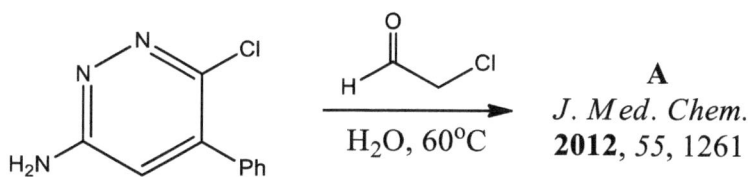

A
J. Med. Chem.
2012, *55*, 1261

H_2O, 60°C

5)

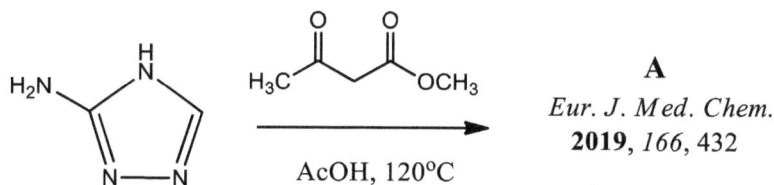

A
Eur. J. Med. Chem.
2019, *166*, 432

AcOH, 120°C

6)

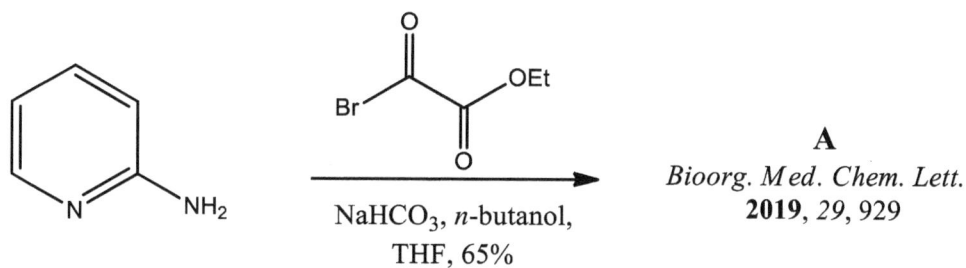

A
Bioorg. Med. Chem. Lett.
2019, *29*, 929

$NaHCO_3$, *n*-butanol,
THF, 65%

7)

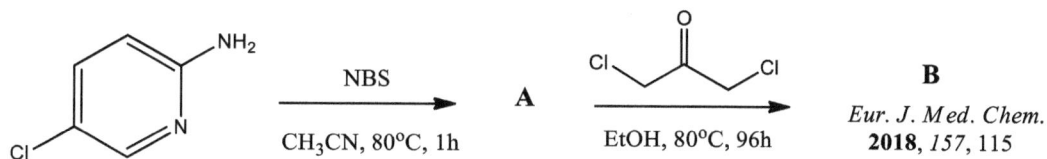

NBS

A

B
Eur. J. Med. Chem.
2018, *157*, 115

CH_3CN, 80°C, 1h

EtOH, 80°C, 96h

8)

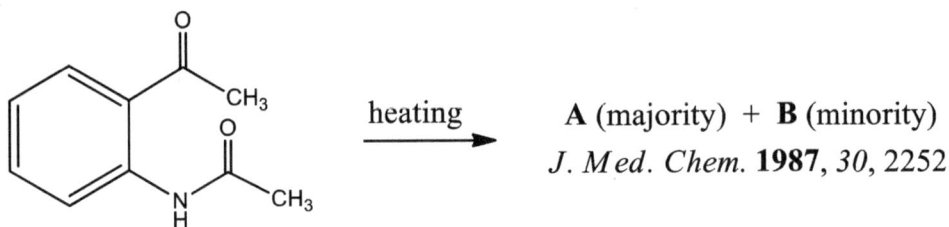

heating → **A** (majority) + **B** (minority)
J. Med. Chem. **1987**, *30*, 2252

9)

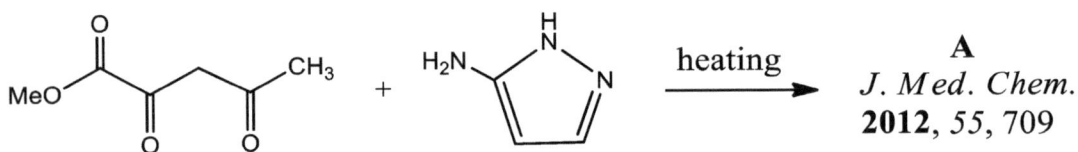

+   heating → **A**
J. Med. Chem.
2012, *55*, 709

10)

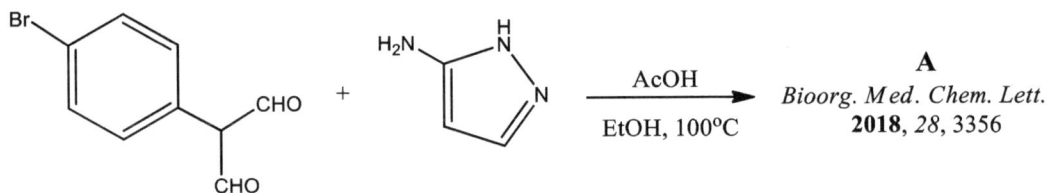

+   AcOH
EtOH, 100°C → **A**
Bioorg. Med. Chem. Lett.
2018, *28*, 3356

11)

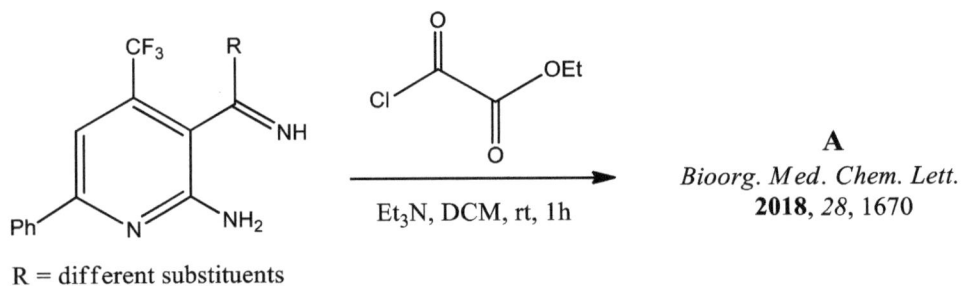

Et₃N, DCM, rt, 1h → **A**
Bioorg. Med. Chem. Lett.
2018, *28*, 1670

R = different substituents

12)

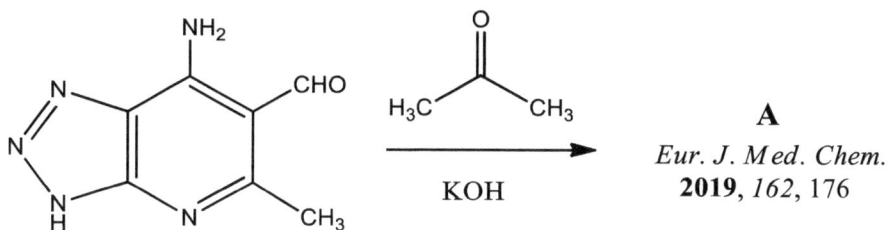

A
Eur. J. Med. Chem.
2019, *162*, 176

13)

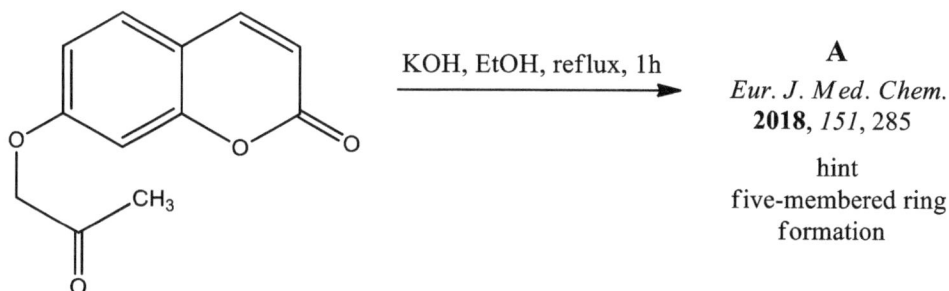

KOH, EtOH, reflux, 1h

A
Eur. J. Med. Chem.
2018, *151*, 285

hint
five-membered ring
formation

14)

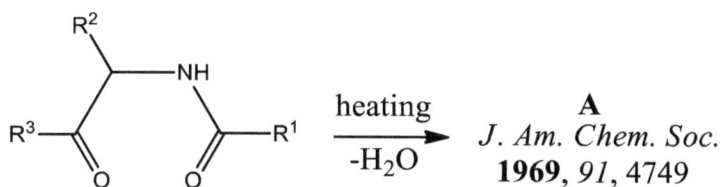

heating
$-H_2O$

A
J. Am. Chem. Soc.
1969, *91*, 4749

15)

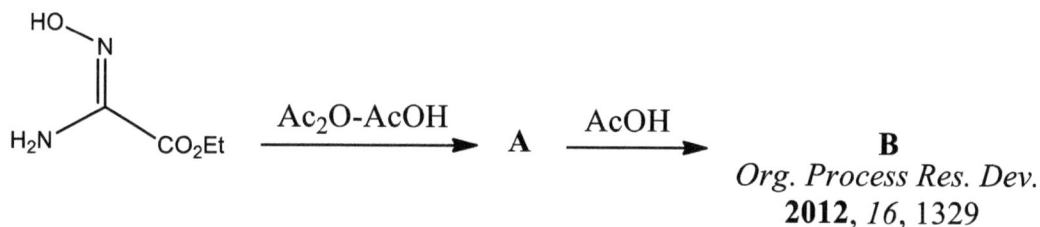

Ac_2O-AcOH   A   AcOH

B
Org. Process Res. Dev.
2012, *16*, 1329

16)

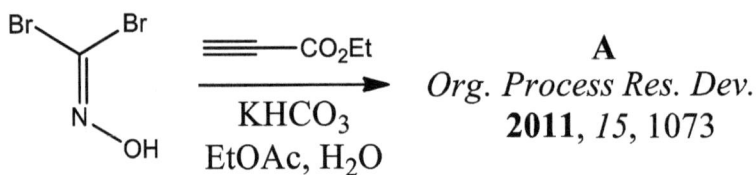

KHCO$_3$

EtOAc, H$_2$O

A
Org. Process Res. Dev.
2011, *15*, 1073

17)

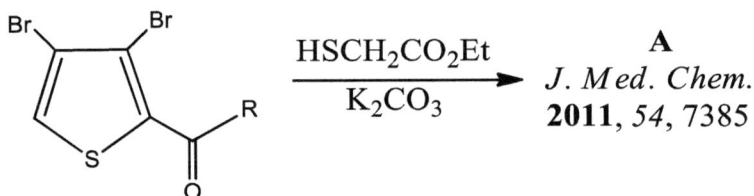

HSCH$_2$CO$_2$Et
――――――――
K$_2$CO$_3$

A
J. Med. Chem.
2011, *54*, 7385

18)

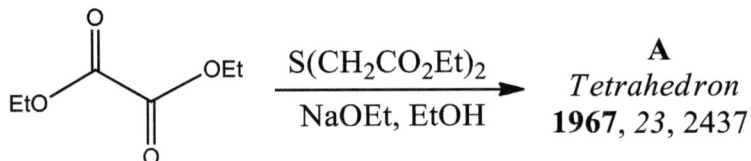

S(CH$_2$CO$_2$Et)$_2$
――――――――
NaOEt, EtOH

A
Tetrahedron
1967, *23*, 2437

19) Proposing the mechanism for the formation of the pyrimidinone nucleus.

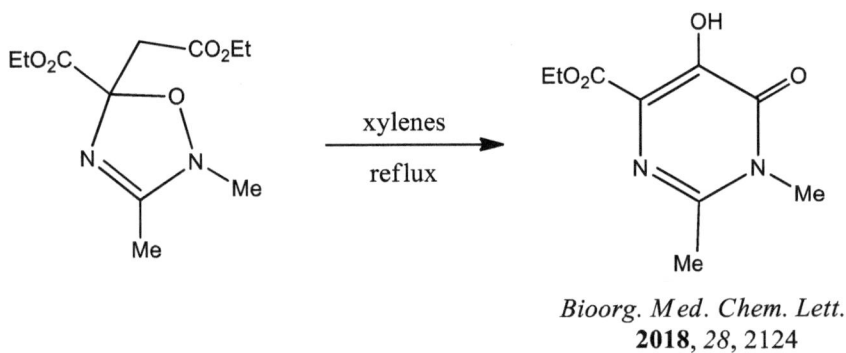

xylenes

reflux

Bioorg. Med. Chem. Lett.
2018, *28*, 2124

20)

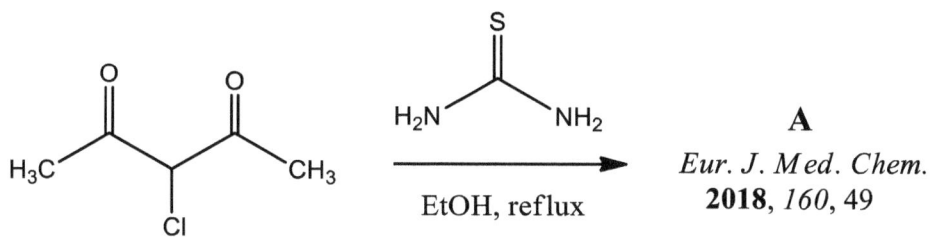

A

Eur. J. Med. Chem.
2018, *160*, 49

21)

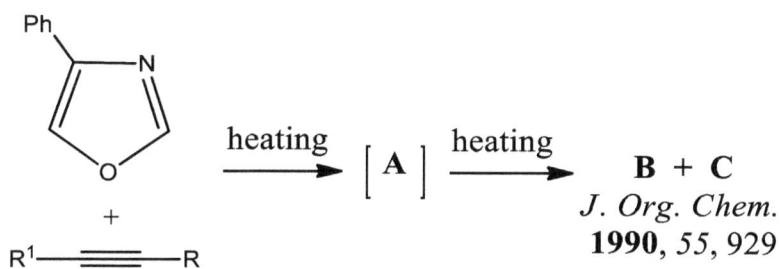

[A]   →   **B + C**
J. Org. Chem.
1990, *55*, 929

22)

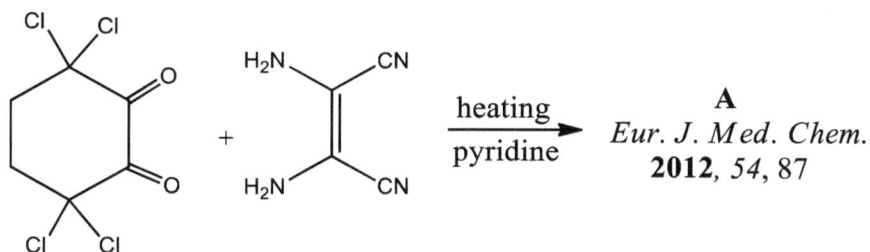

A
Eur. J. Med. Chem.
2012, *54*, 87

Answers

1)

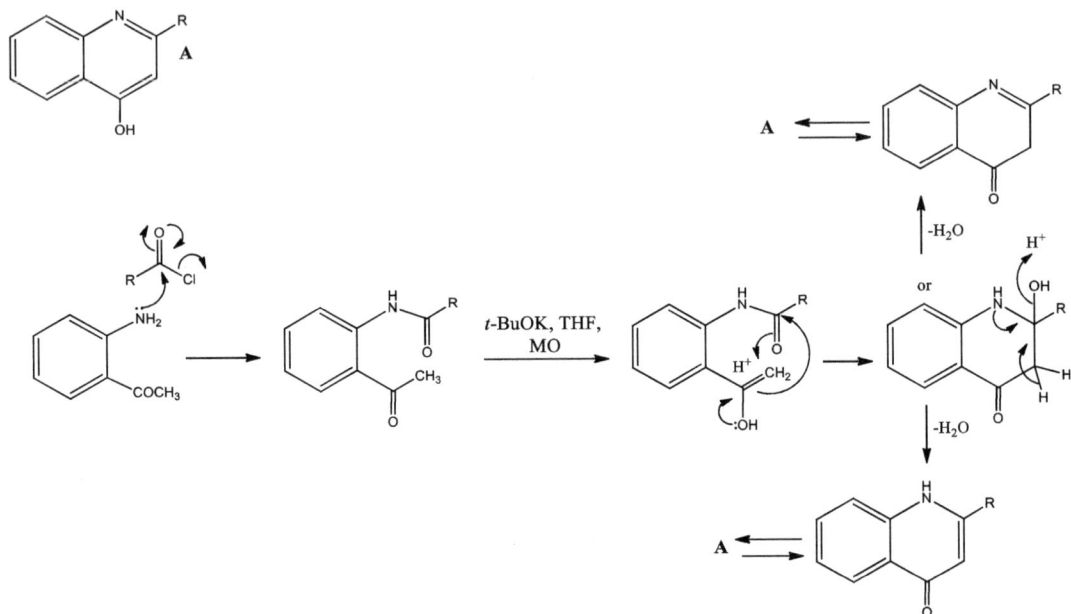

2)

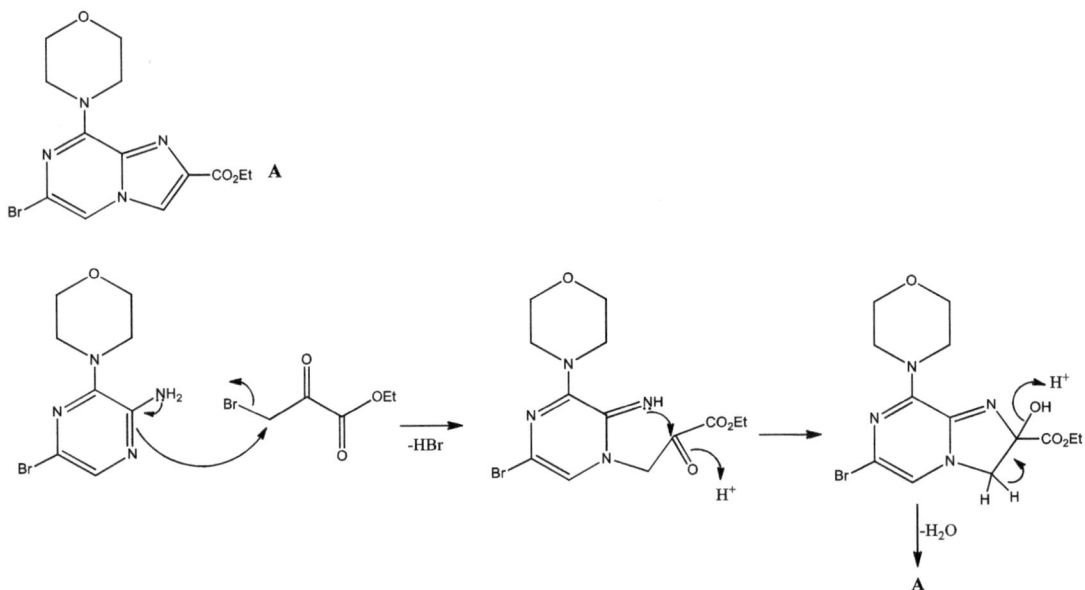

3)

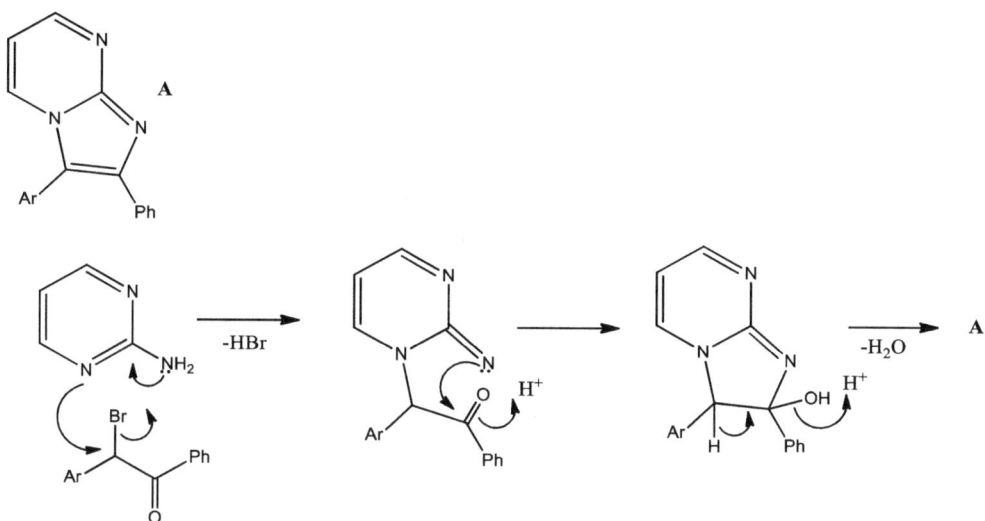

4)

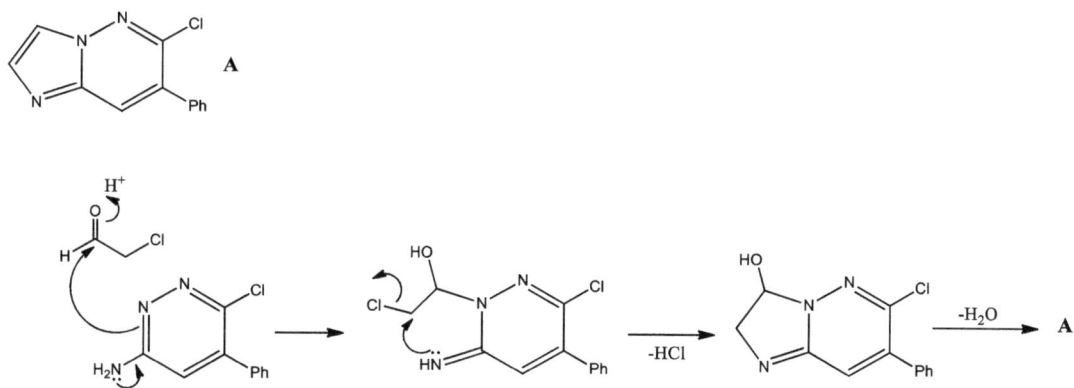

5)

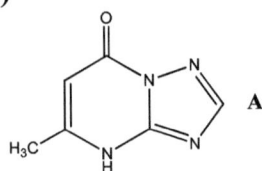

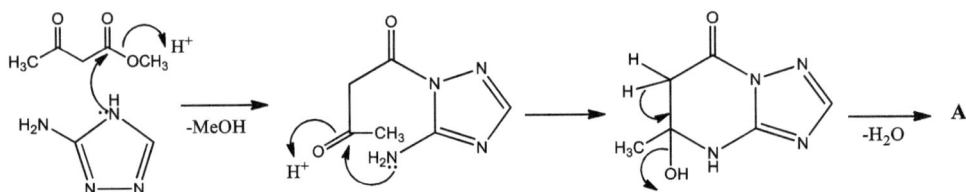

6)

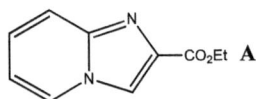

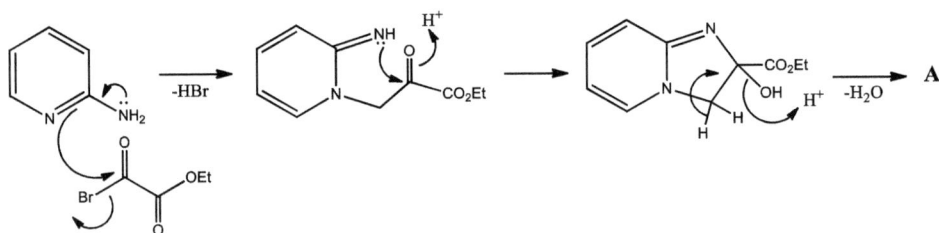

7)

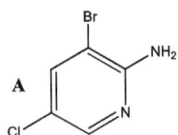

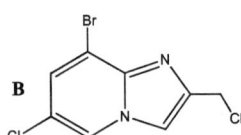

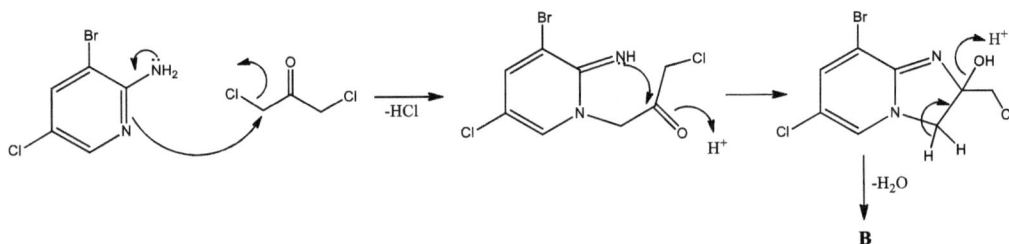

8)

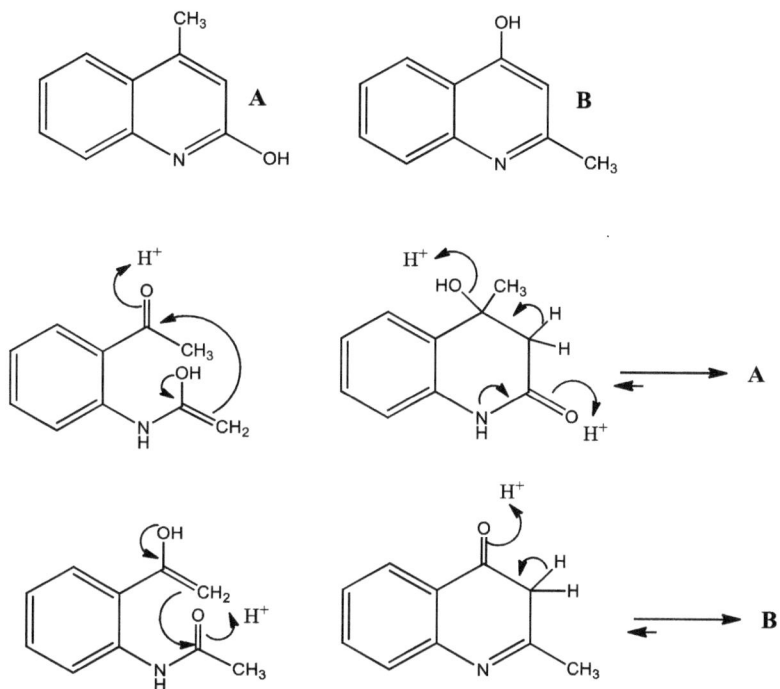

9)

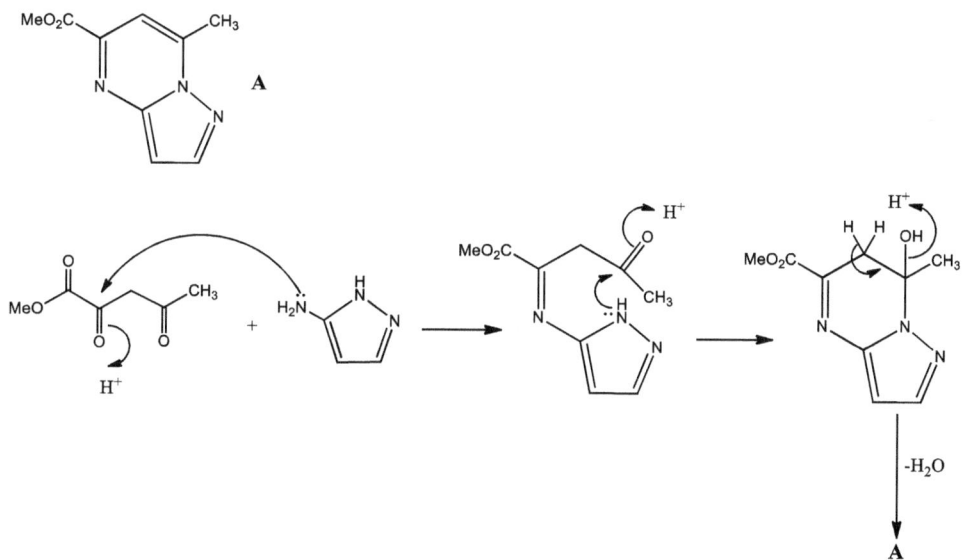

10)

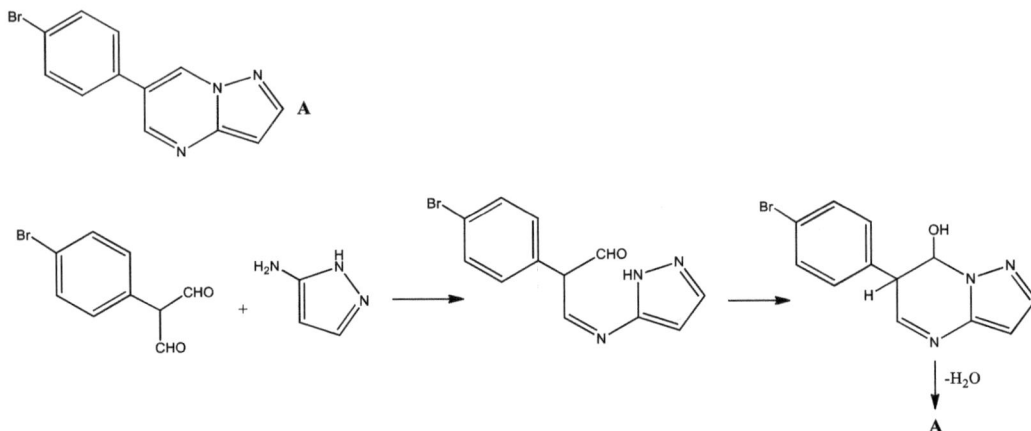

11)

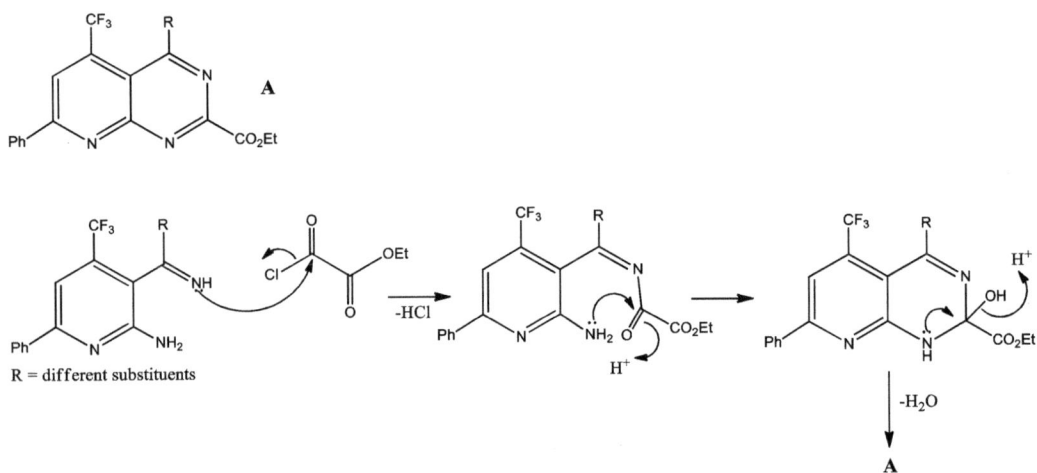

R = different substituents

12)

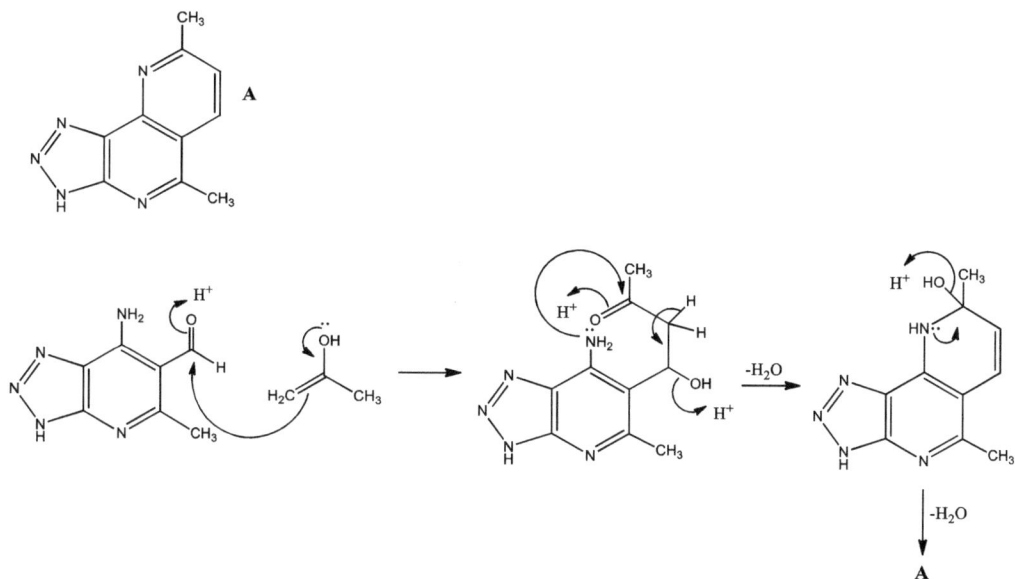

13)

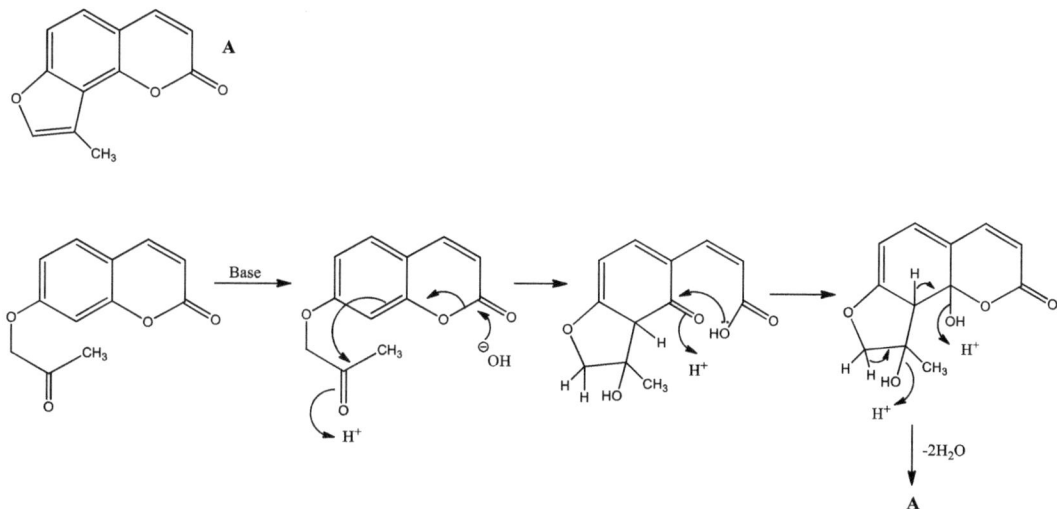

14)

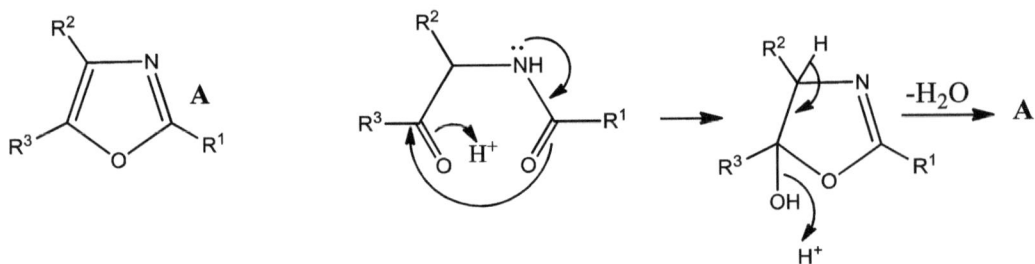

15)

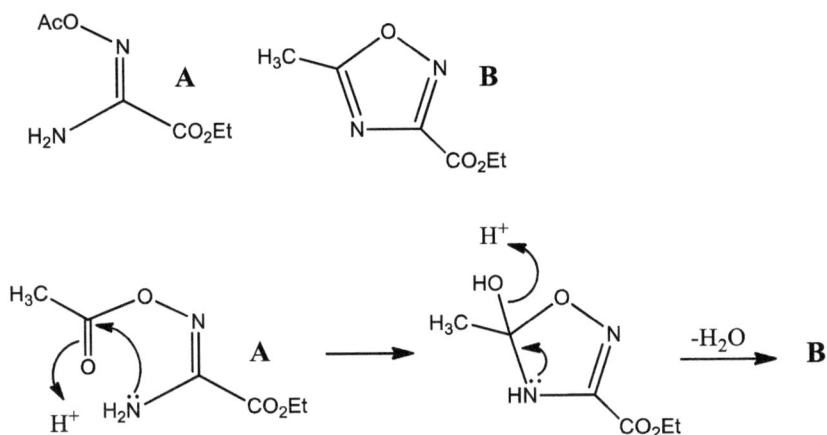

16)

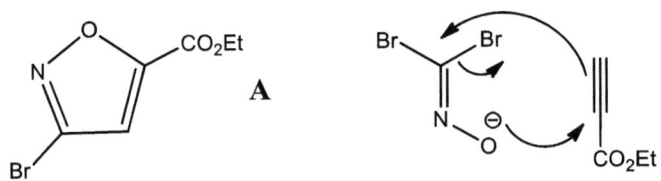

17)

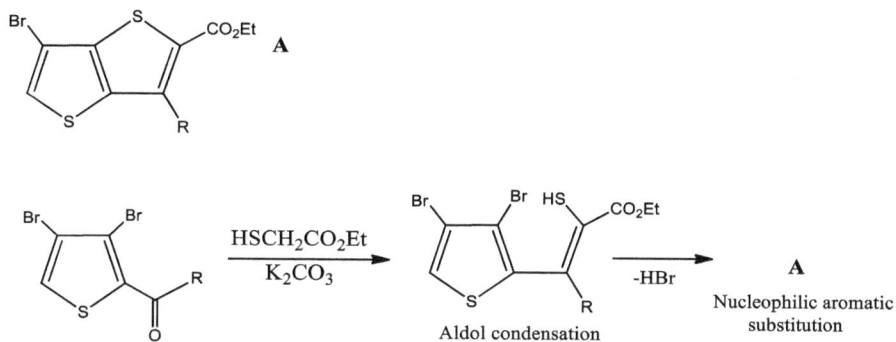

18)

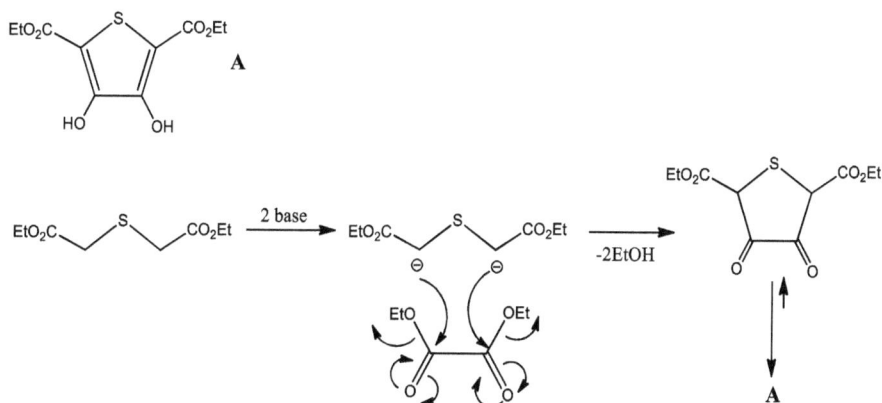

19)

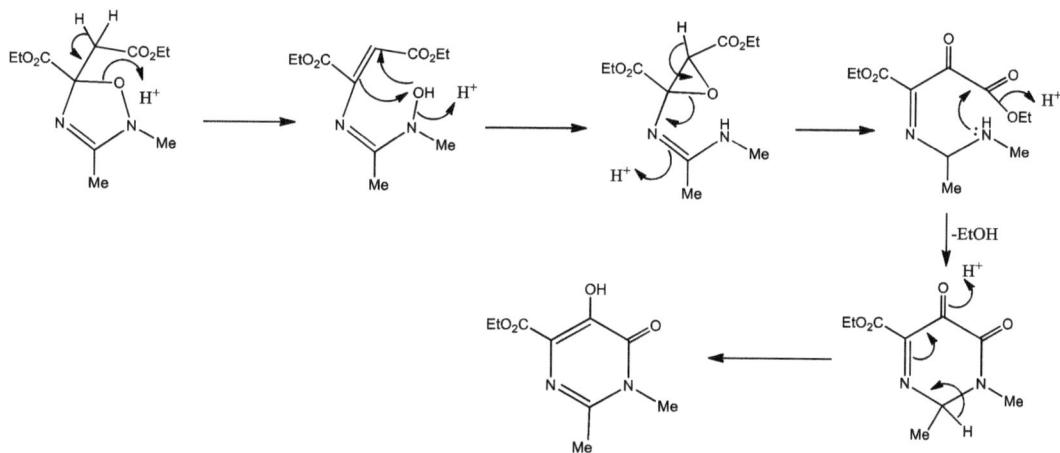

20)

This exercise exemplifies well what we mentioned in the example. We can observe that, although the reagents used are different, the mechanism is very similar, emphasizing again the need to understand the mechanisms and not to decorate them.

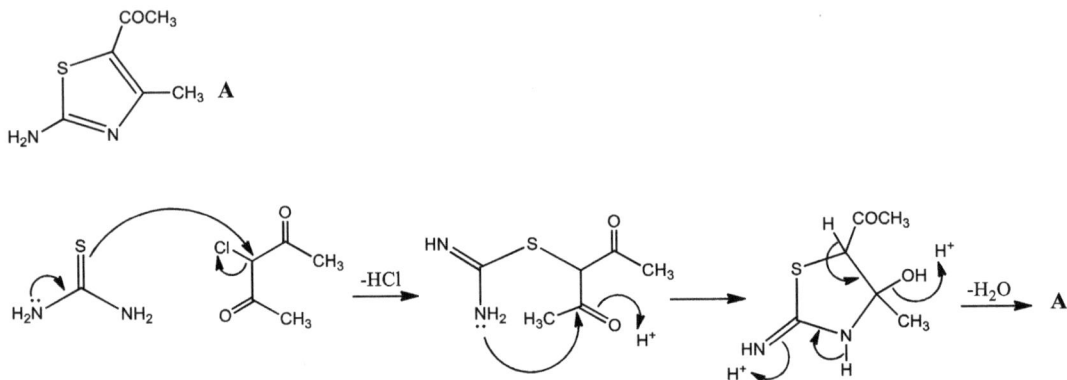

21)

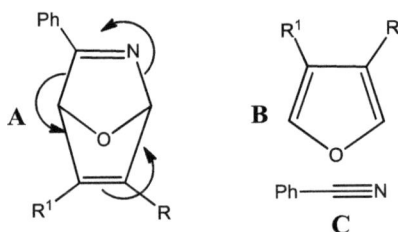

22)

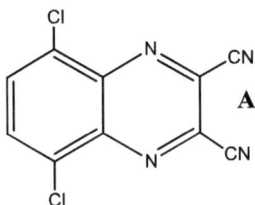

CONCLUSION

Heteroaromatics are fundamental in developing new drugs, being present in many drugs and bioactive compounds. Due to their importance and immense structural variety, this chapter's exercises are based on the formation of different heteroaromatic substances. Therefore, the reader will become familiar with the most common nuclei and their synthesis. Besides, with the exercises, the reader will also be able to propose changes in their structures.

CHAPTER 5

Hydrazine and its Derivatives

Abstract: The chemical hydrazine (NH_2NH_2) and its derivatives have several applications being also responsible for a large number of chemical transformations. This class has applications as rocket fuel and being present in the composition of explosives, in the control of corrosion, and the area of polymers. They also have full use in the development of bioactive substances being of great importance in medicinal chemistry. In the exercises of this chapter, we can synthesize several rings using hydrazine and its derivatives, and it is up to the readers to identify them.

Keywords: Bioactive compounds, Chemical transformations, Drugs, Exercises, Heteroaromatic compounds, Hydrazine, Medicinal chemistry, Molecules, Reaction conditions, Reagents, Retrosynthetic analysis, Substances, Synthesis, Synthons.

INTRODUCTION

The chemical hydrazine (NH_2NH_2) and its derivatives have several applications, being also responsible for a large number of chemical transformations. This class has applications as rocket fuel and being present in the composition of explosives, in the control of corrosion, and the area of polymers. They also have full use in the development of bioactive substances being of great importance in medicinal chemistry. Some examples of hydrazine derivatives such as, methylhydrazine, phenylhydrazine, *N*-acylhydrazide, semicarbazide, and thiosemicarbazide are listed below. It is noteworthy that the condensation of these substances above listed with aldehydes produces the respective hydrazones, methylhydrazones, phenylhydrazones, *N*-acylhydrazones, semicarbazones, and thiosemicarbazones (Scheme **1**).

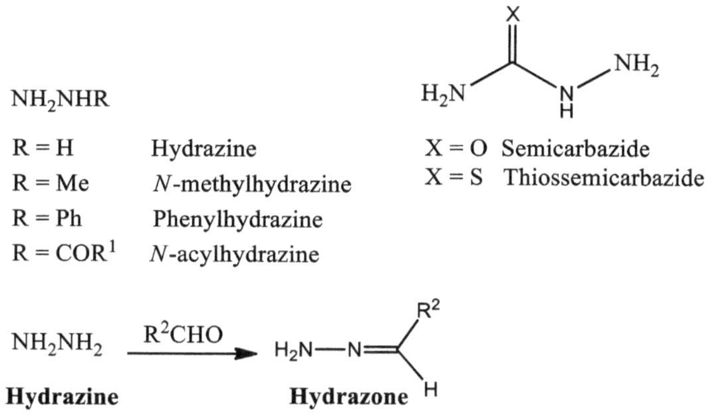

NH$_2$NHR

R = H   Hydrazine
R = Me   *N*-methylhydrazine
R = Ph   Phenylhydrazine
R = COR^1 *N*-acylhydrazine

X = O Semicarbazide
X = S Thiossemicarbazide

NH$_2$NH$_2$ →   H$_2$N—N=

Hydrazine   **Hydrazone**

Scheme 1. Hydrazine and its derivatives.

Example

After the reaction with the hydrazine (NH$_2$NH$_2$), what will be the target molecule formed?

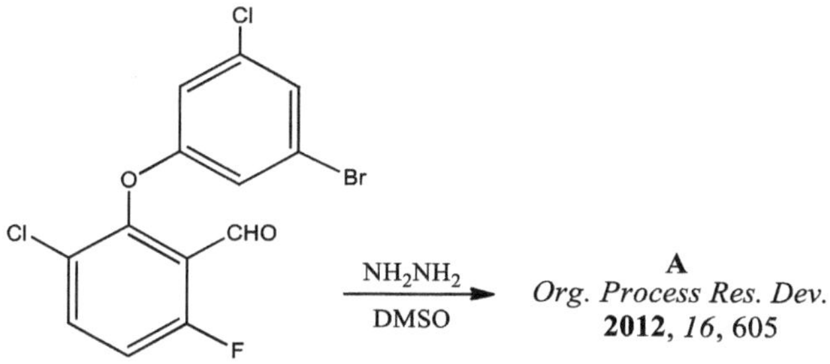

Answer

The use of hydrazine was responsible for the formation of the indazole nucleus due to the position of the aldehyde chemical function (hydrazone formation) and the fluorine atom acting as a leaving group.

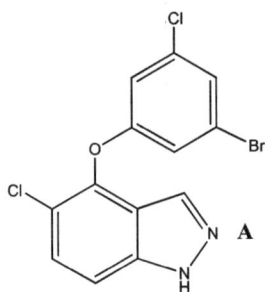

In the exercises below, we can synthesize several rings using hydrazine and its derivatives, and it is up to the reader to identify them.

EXERCISES

1)

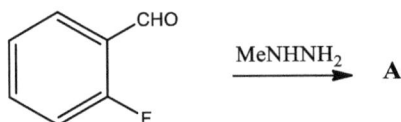

Org. Process Res. Dev.
2011, *15*, 565

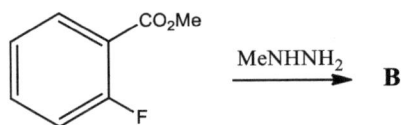

2)

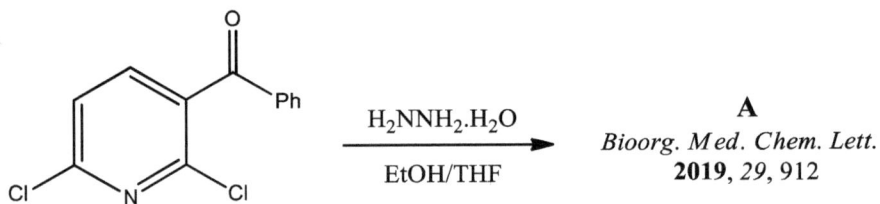

3)

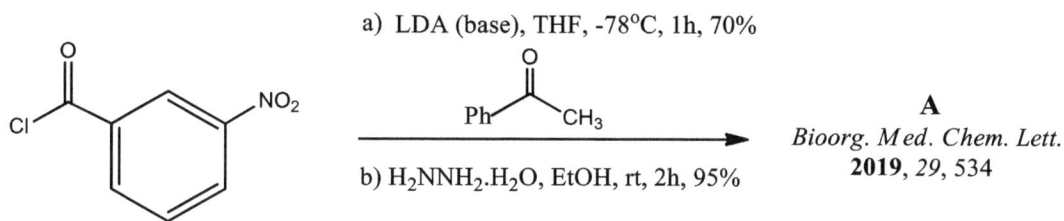

a) LDA (base), THF, -78°C, 1h, 70%

b) H$_2$NNH$_2$.H$_2$O, EtOH, rt, 2h, 95%

A
Bioorg. Med. Chem. Lett.
2019, *29*, 534

4)

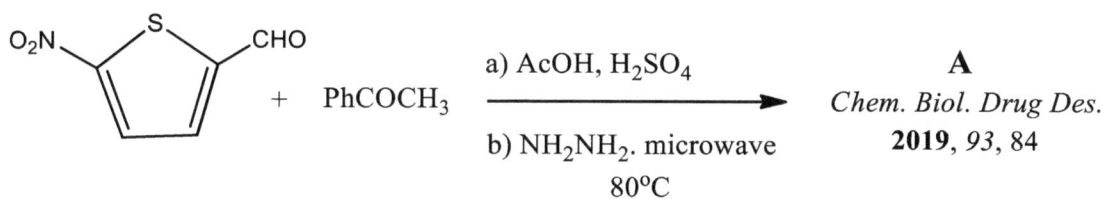

+ PhCOCH$_3$

a) AcOH, H$_2$SO$_4$

b) NH$_2$NH$_2$. microwave 80°C

A
Chem. Biol. Drug Des.
2019, *93*, 84

5)

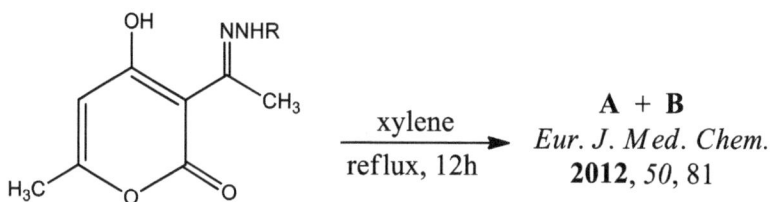

xylene
reflux, 12h

A + B
Eur. J. Med. Chem.
2012, *50*, 81

6)

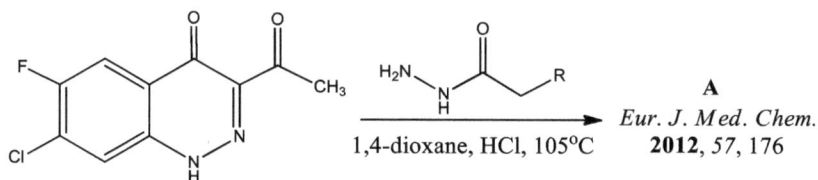

1,4-dioxane, HCl, 105°C

A
Eur. J. Med. Chem.
2012, *57*, 176

7)

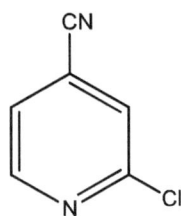

a) $NH_2NH_2.H_2O$, 1-butanol/THF
60-100°C

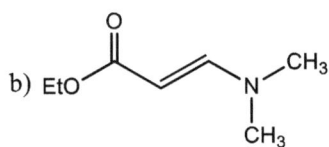

b)

AcOH, EtOH, 90°C

c) *t*-BuOK, EtOH, rt

Hint: -$N(CH_3)_2$ is a good leaving group

A
Bioorg. Med. Chem. Lett.
2018, *28*, 1490

8)

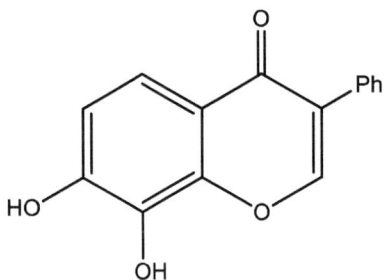

$NH_2NH_2.H_2O$

THF

A
Eur. J. Med. Chem.
2018, *158*, 707

9)

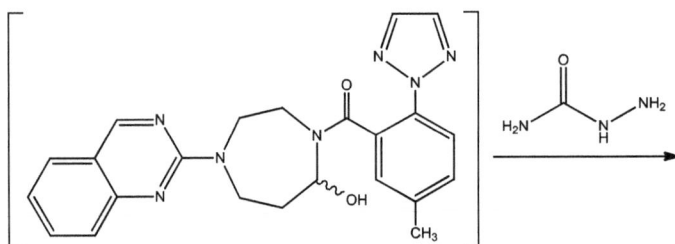

A
Bioorg. Med. Chem. Lett.
2018, *28*, 3283

10)

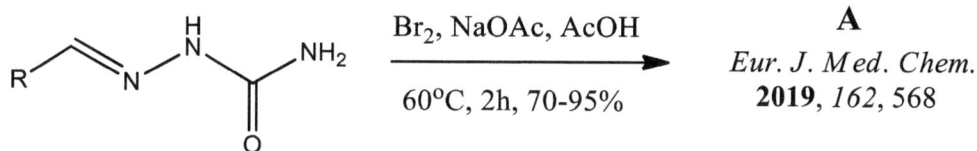

$$\xrightarrow[\text{60°C, 2h, 70-95\%}]{\text{Br}_2,\ \text{NaOAc, AcOH}}$$

A
Eur. J. Med. Chem.
2019, *162*, 568

R = different substituents

11)

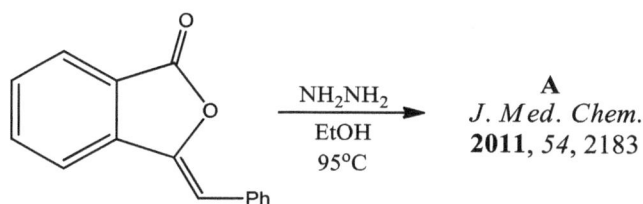

$$\xrightarrow[\substack{\text{EtOH}\\95°\text{C}}]{\text{NH}_2\text{NH}_2}$$

A
J. Med. Chem.
2011, *54*, 2183

12)

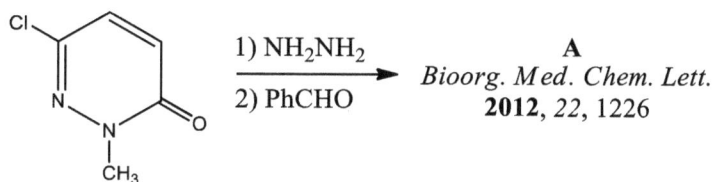

$$\xrightarrow[\text{2) PhCHO}]{\text{1) NH}_2\text{NH}_2}$$

A
Bioorg. Med. Chem. Lett.
2012, *22*, 1226

13)

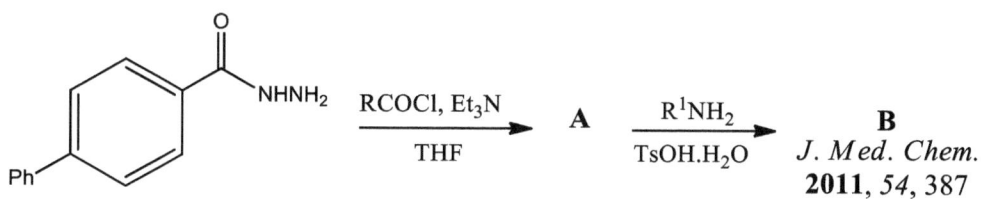

$$\xrightarrow[\text{THF}]{\text{RCOCl, Et}_3\text{N}}$$ **A** $$\xrightarrow[\text{TsOH.H}_2\text{O}]{\text{R}^1\text{NH}_2}$$

B
J. Med. Chem.
2011, *54*, 387

14)

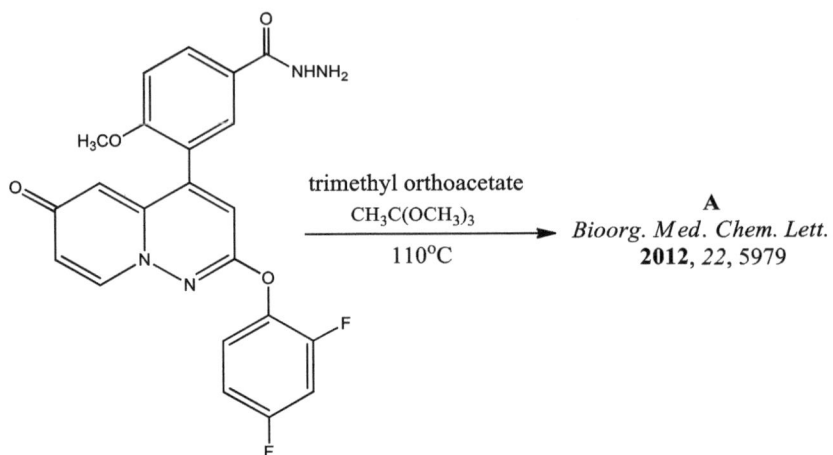

trimethyl orthoacetate
CH$_3$C(OCH$_3$)$_3$
110°C

A
Bioorg. Med. Chem. Lett.
2012, *22*, 5979

15)

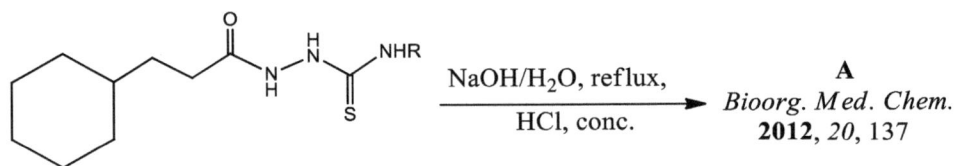

NaOH/H$_2$O, reflux,
HCl, conc.

A
Bioorg. Med. Chem.
2012, *20*, 137

16)

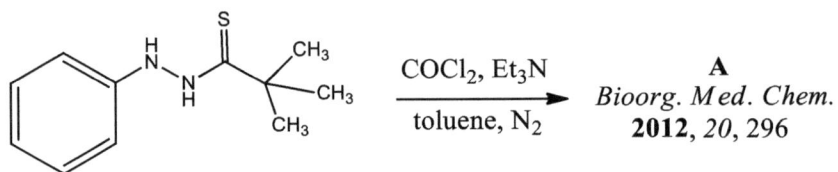

COCl$_2$, Et$_3$N
toluene, N$_2$

A
Bioorg. Med. Chem.
2012, *20*, 296

Hint: The first step of the reaction occurs in one of the nitrogen of the start material.

17)

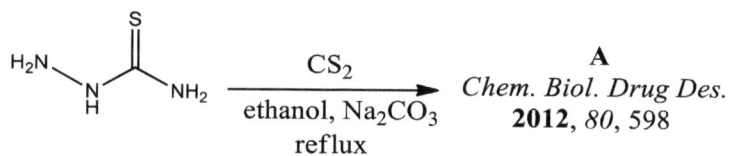

H_2N–NH–C(=S)–NH_2 → $\xrightarrow[\text{reflux}]{\substack{CS_2 \\ \text{ethanol, } Na_2CO_3}}$ → **A**
Chem. Biol. Drug Des.
2012, *80*, 598

18)

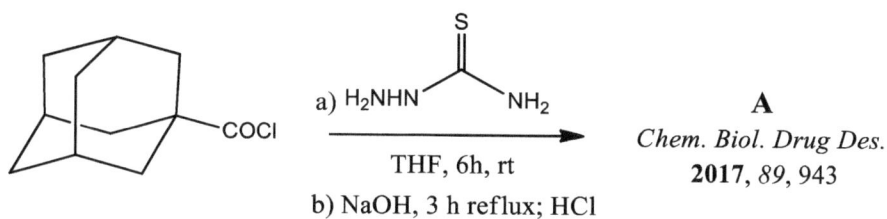

a) H_2NHN–C(=S)–NH_2
$\xrightarrow{\text{THF, 6h, rt}}$
b) NaOH, 3 h reflux; HCl

A
Chem. Biol. Drug Des.
2017, *89*, 943

19)

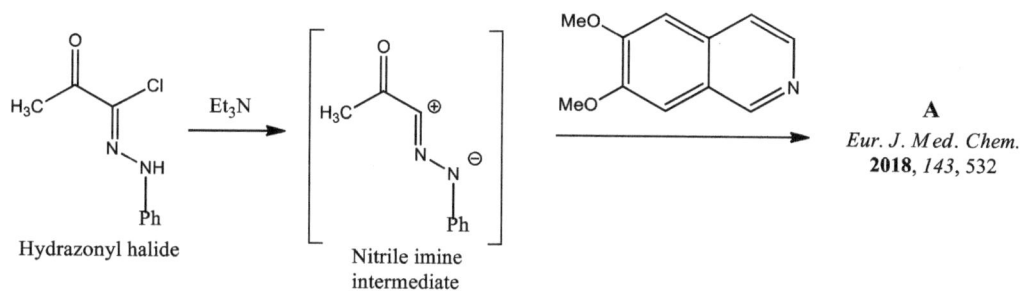

$\xrightarrow{Et_3N}$

Hydrazonyl halide

Nitrile imine
intermediate

MeO / MeO isoquinoline

A
Eur. J. Med. Chem.
2018, *143*, 532

20)

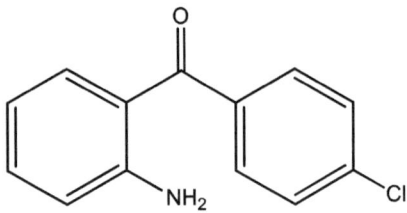

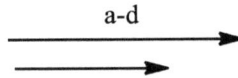

A
Chem. Biol. Drug Des.
2017, *90*, 97

a) 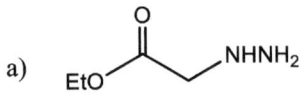 EtOH, microwave, 130 °C, 2.5 h; b) K$_2$CO$_3$, MeCN, 25°C, 12 h

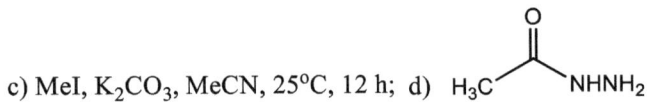

c) MeI, K$_2$CO$_3$, MeCN, 25°C, 12 h; d) 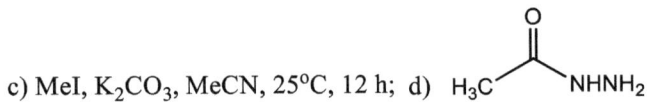 H$_3$C NHNH$_2$

Answer

1)

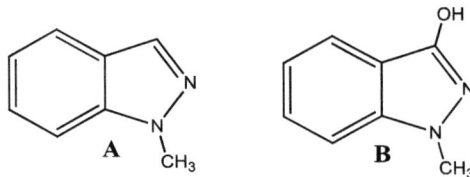

2)

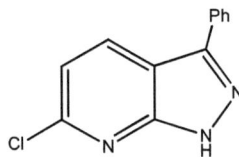

3)

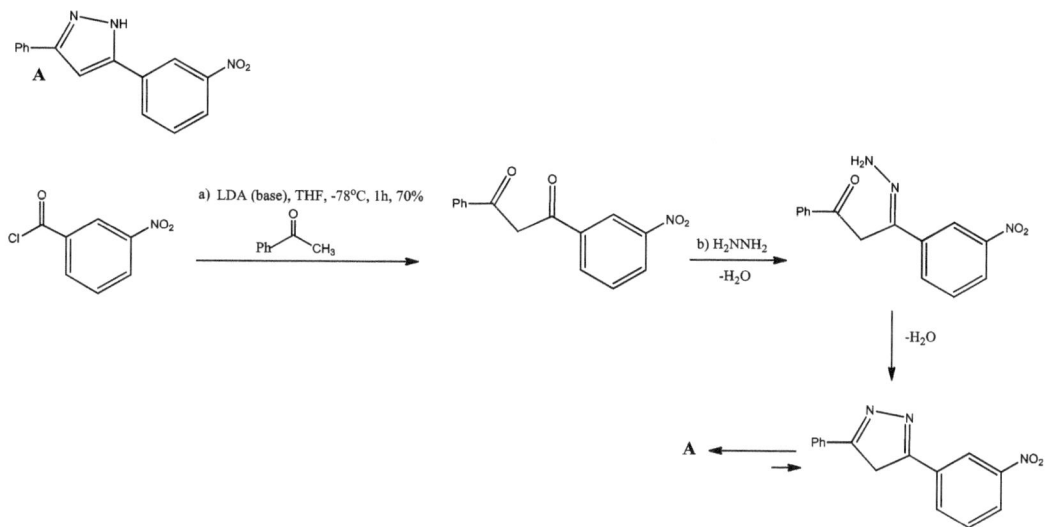

4)

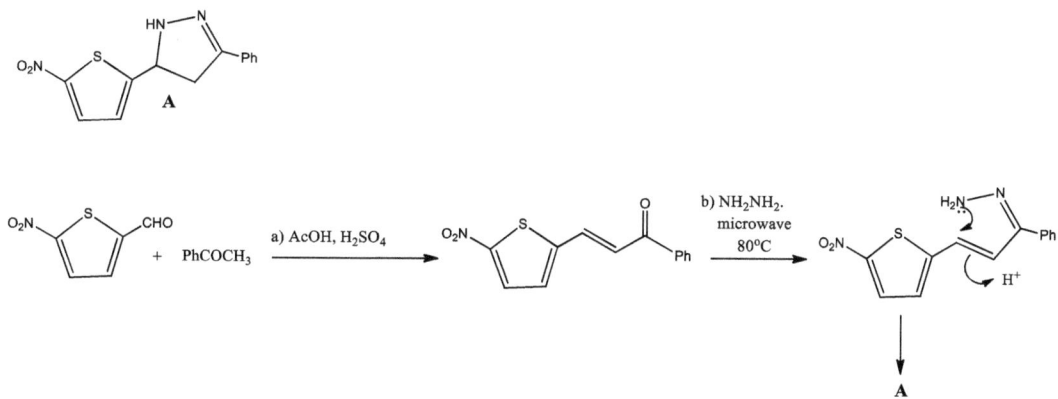

5)

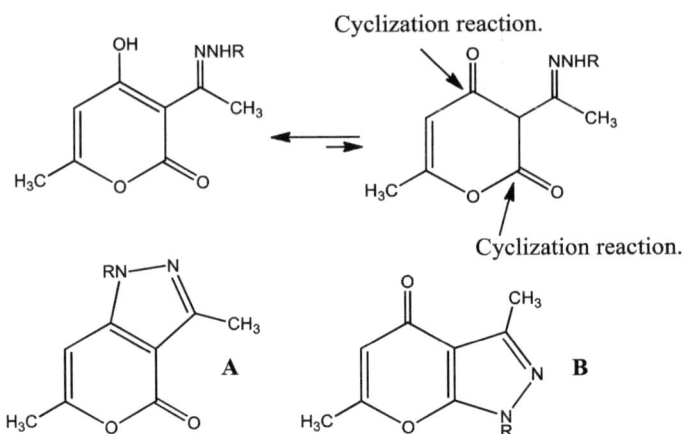

6)

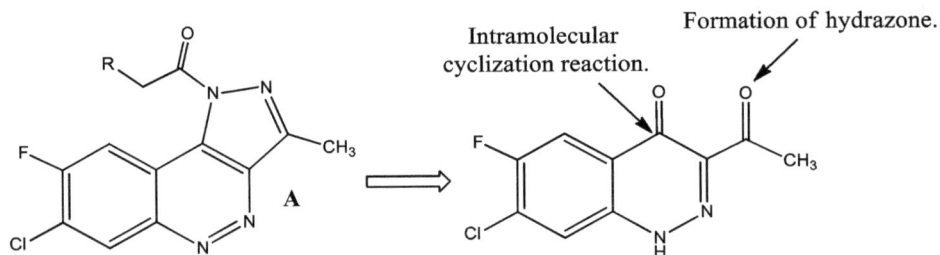

7)

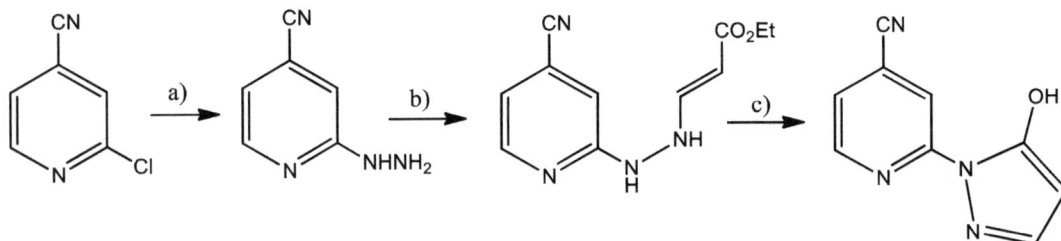

8)

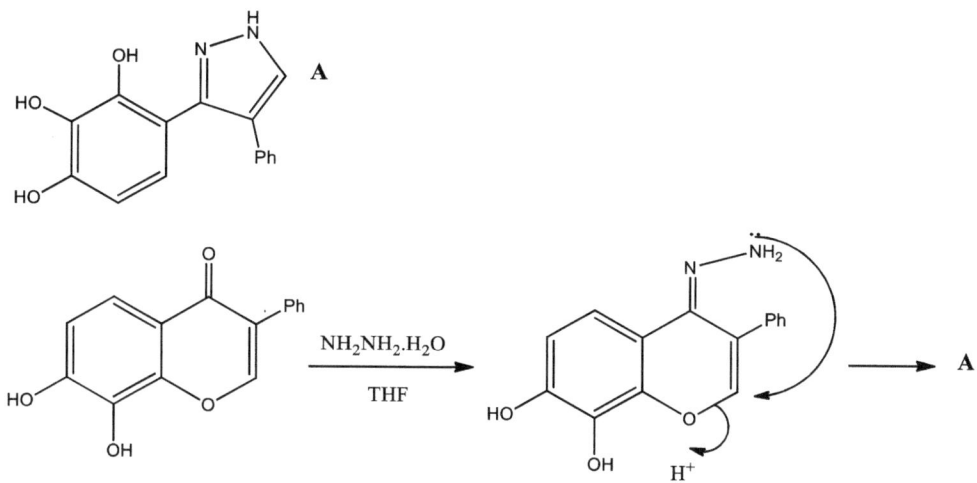

9)

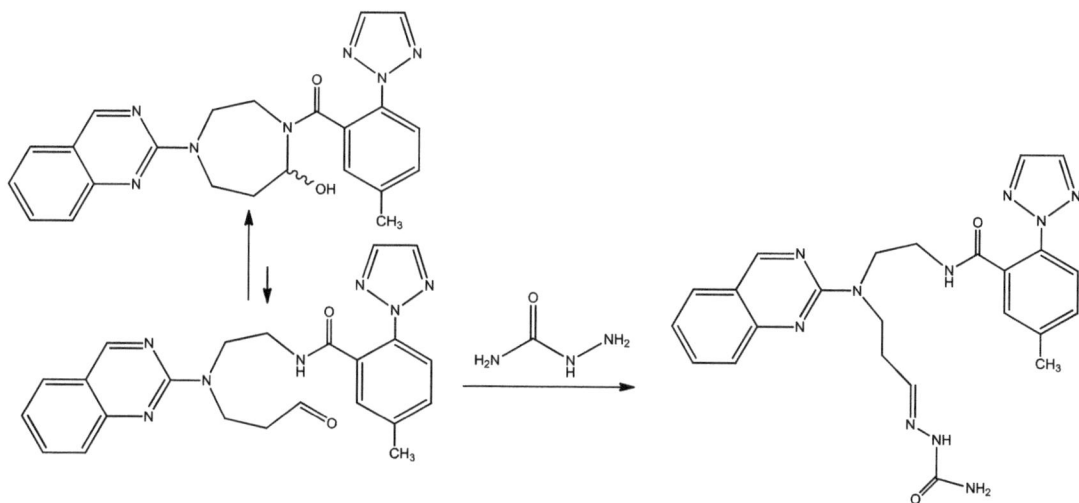

10)

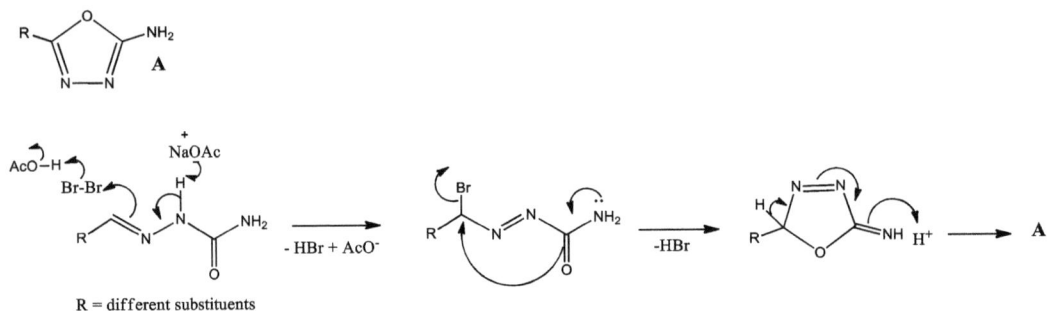

R = different substituents

11)

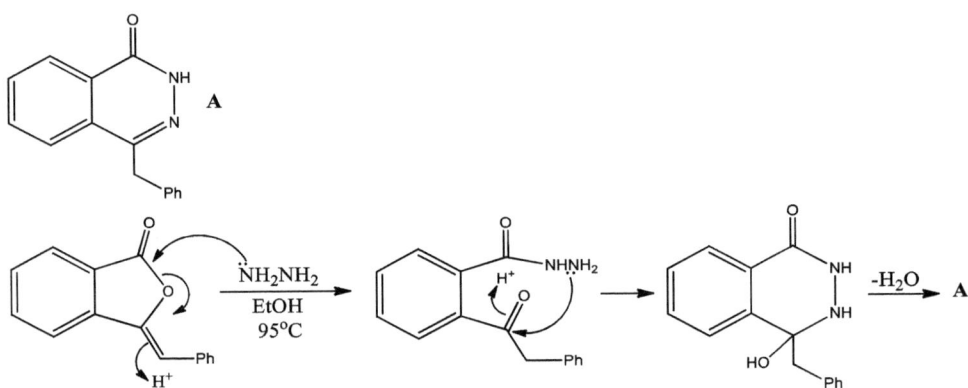

12)

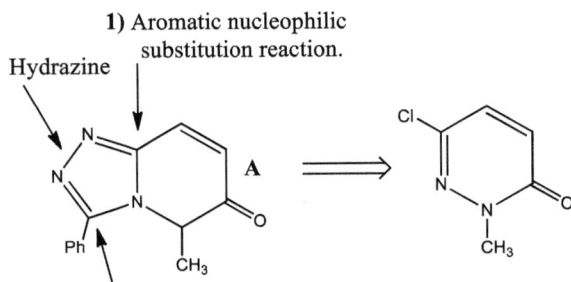

1) Aromatic nucleophilic substitution reaction.

Hydrazine

2) Benzaldehyde. Nucleophilic addition reaction to carbonyl followed by the reaction of cyclization and aromatization of the system.

13)

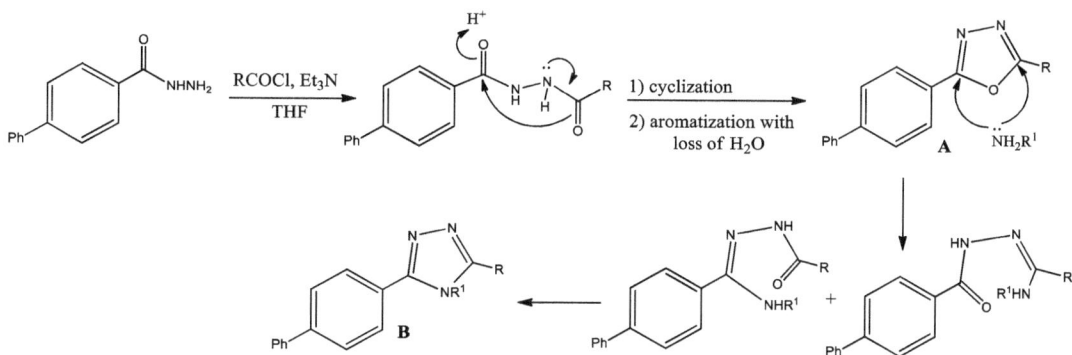

14)

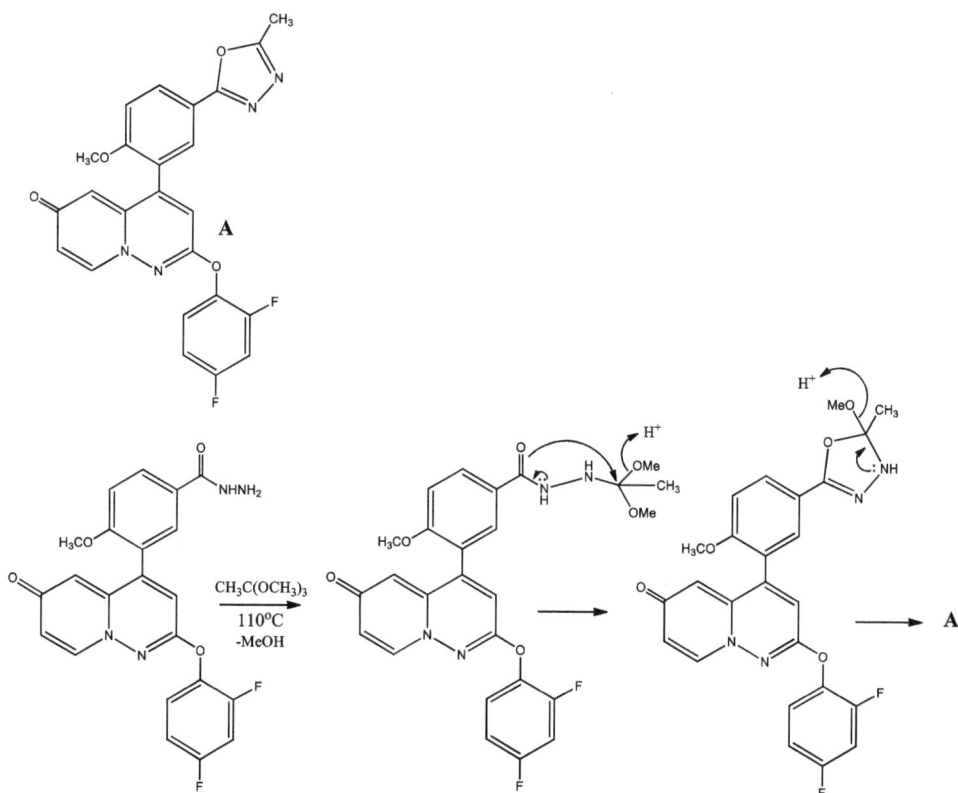

15)

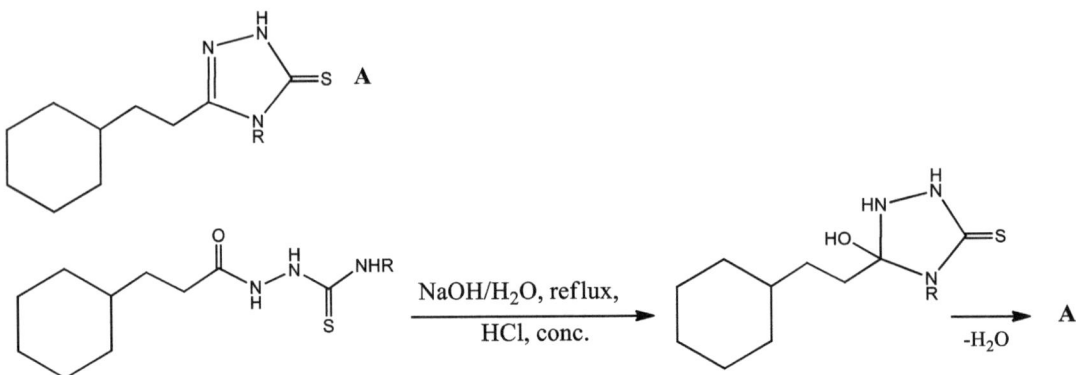

16)

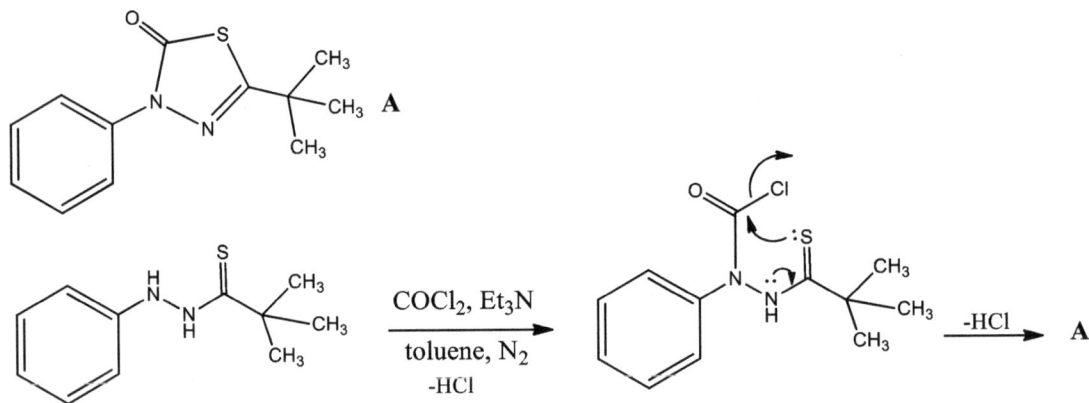

17)

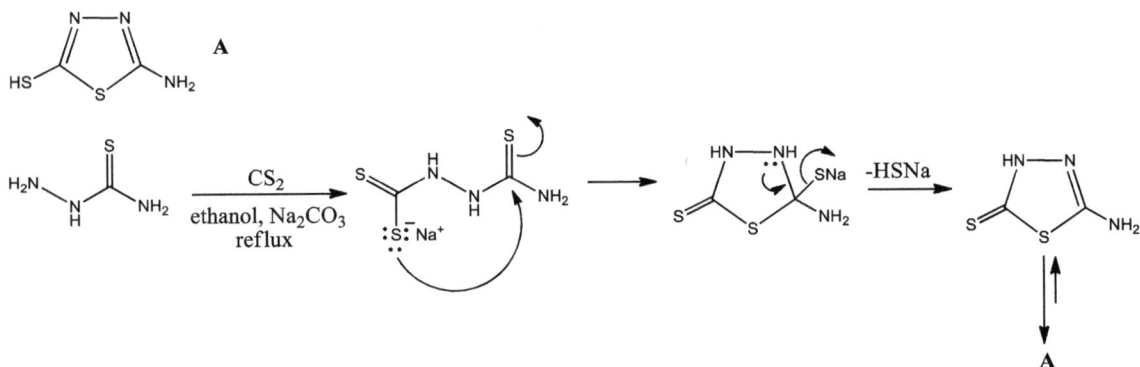

18)

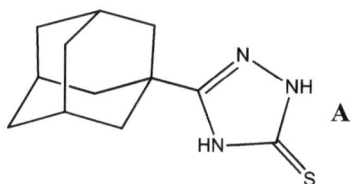

1,2,4- triazole-3-thione nucleus

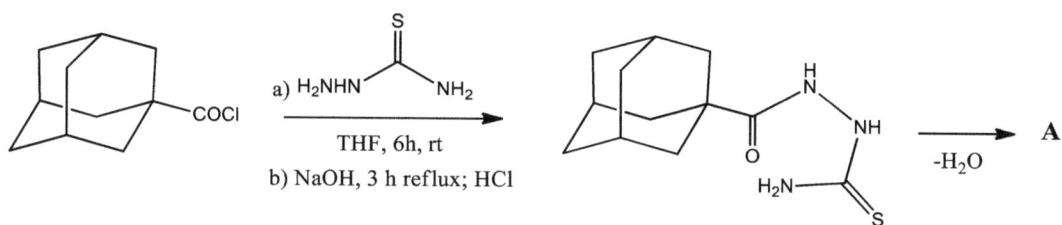

19)

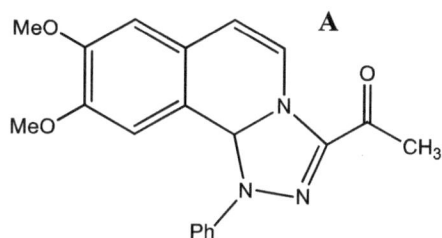

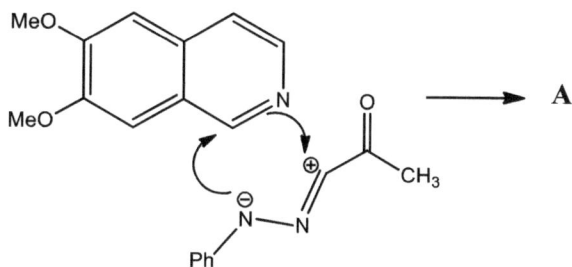

20)

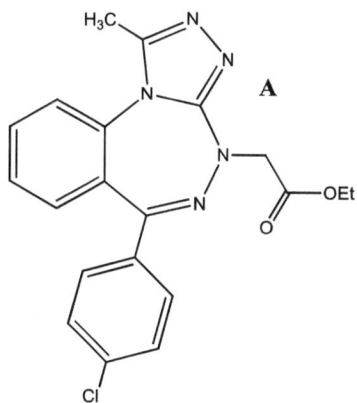

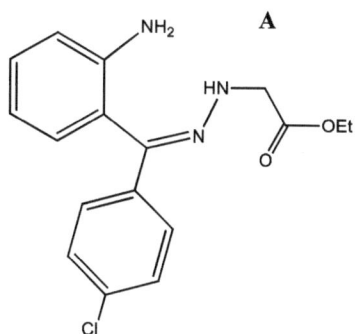

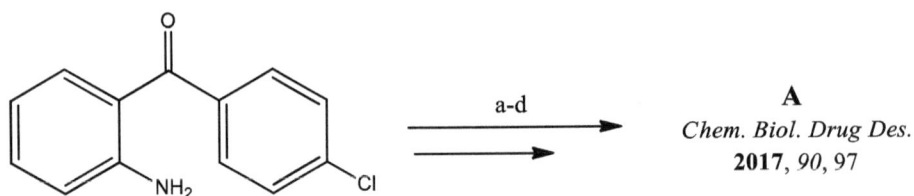

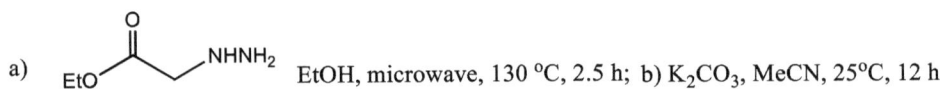

a-d → **A**
Chem. Biol. Drug Des.
2017, *90*, 97

a) EtOH, microwave, 130 °C, 2.5 h; b) K_2CO_3, MeCN, 25°C, 12 h

c) MeI, K_2CO_3, MeCN, 25°C, 12 h; d)

CONCLUSION

The chemical hydrazine (NH_2NH_2) and its derivatives have several applications in the development of bioactive substances being of great importance in medicinal chemistry. Its applications can be seen in the formation of different nuclei and a large number of chemical transformations. The exercises in this chapter demonstrate the versatility of this chemical class as an essential tool in the construction of different types of nuclei.

CHAPTER 6

Nitriles

Abstract: Nitrile is a chemical function found in several drugs in the market for different kinds of diseases and can be transformed into several other functional groups such as ketones, aldehydes, carboxylic acids, amines, amides, and other functional groups. The use of nitriles is also advantageous in forming heteroaromatic nuclei with important applications in medicinal chemistry. In this chapter, the exercises are based on the use of nitriles to form a different nucleus.

Keywords: Bioactive compounds, Chemical transformations, Drugs, Exercises, Heteroaromatic compounds, Hydrazine, Medicinal chemistry, Molecules, Nitriles, Reaction conditions, Reagents, Retrosynthetic analysis, Substances, Synthesis, Synthons.

INTRODUCTION

Nitrile is a chemical function found in medicinal chemistry being able to be transformed in several other functional groups such as tetrazoles, ketones, aldehydes, carboxylic acids, amines, amides, different nucleus and other functional groups (Scheme **1**).

Nitrile is a chemical function found in several drugs in the market for different diseases (Fig. **1**).

Saxagliptin, an oral inhibitor class of dipeptidyl peptidase-4 (DPP-4), used against hypoglycemia. Anastrazole used to treat and prevent breast câncer. Escitalopram used against depression is a selective serotonin reuptake inhibitor. Etravirine is an anti-HIV antiviral being the first of a new class of non-nucleoside inhibitors of the reverse transcriptase (NNRTI), an essential enzyme of HIV. Milrinone is used to treat heart failure acting as a phosphodiesterase inhibitor, and Febuxostat is used to treat gout, working against high uric acid levels.

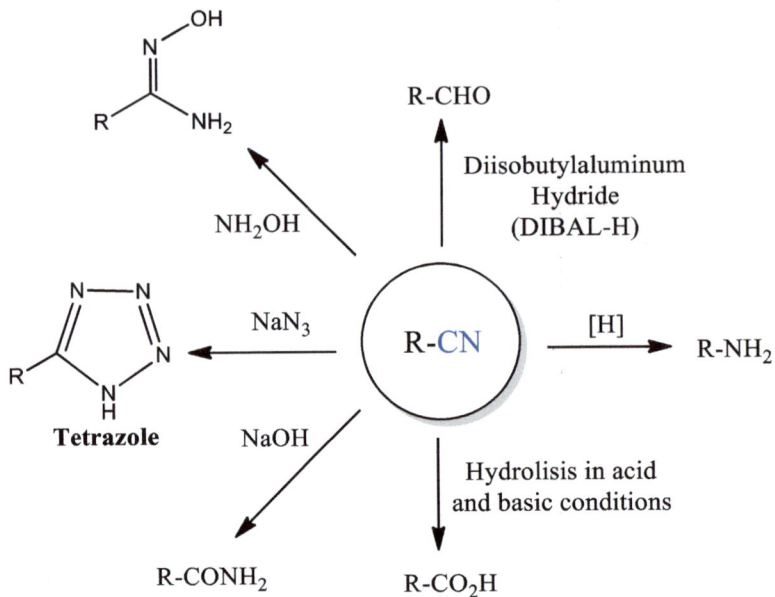

Scheme 1. Chemical transformations of nitrile group.

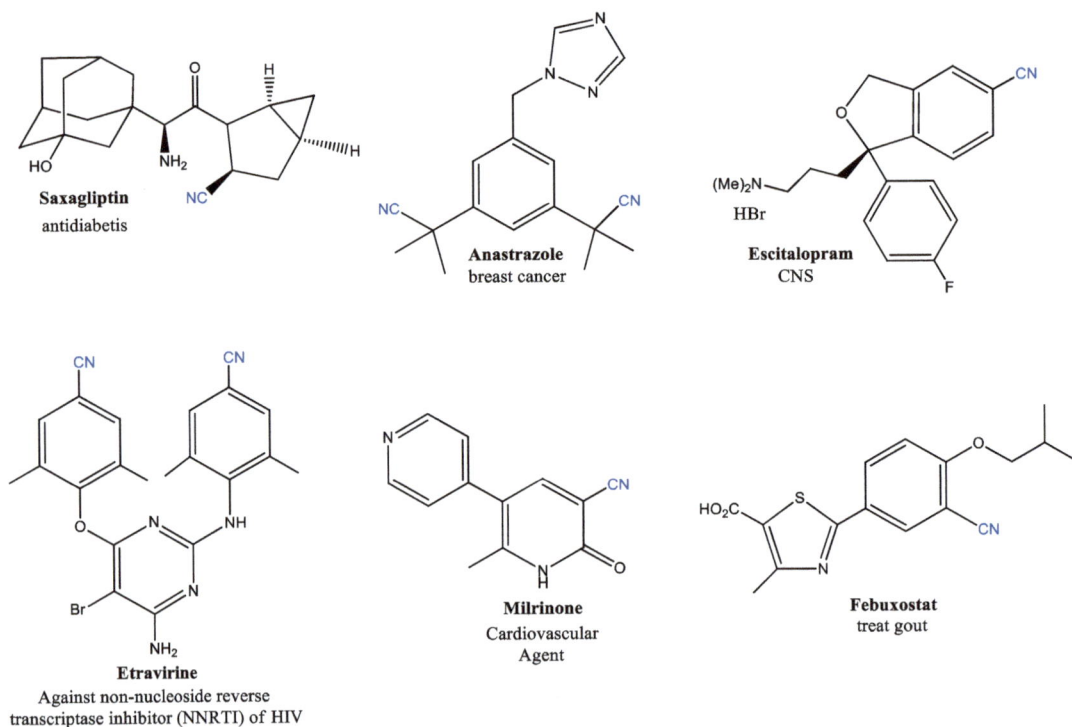

Fig. (1). Drugs containing a nitrile group into its structure.

Example

Which heteroaromatic nucleus is formed using malononitrile and an organic azide?

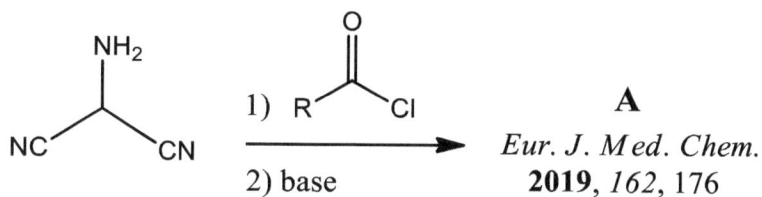

Answer

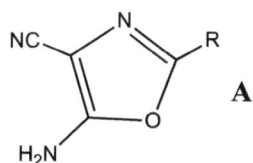

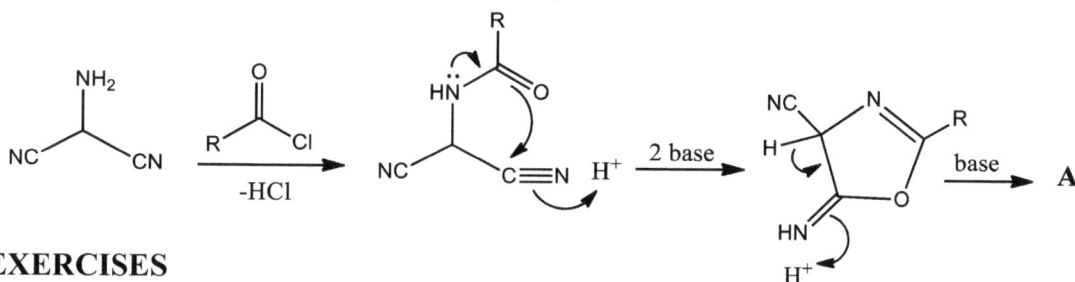

EXERCISES

Based on the use of nitriles, in the reactions below, the heteroaromatic nucleus formed has been identified.

1)

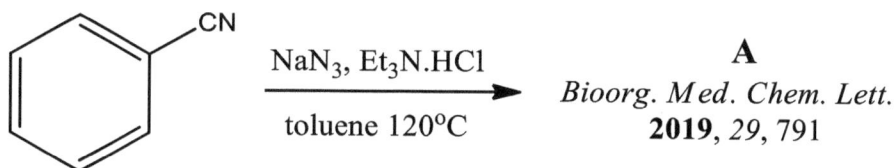

2)

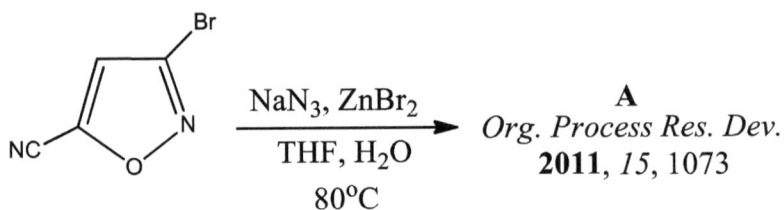

$$\xrightarrow[\substack{\text{THF, H}_2\text{O} \\ 80°C}]{\text{NaN}_3, \text{ZnBr}_2}$$

A
Org. Process Res. Dev.
2011, *15*, 1073

3)

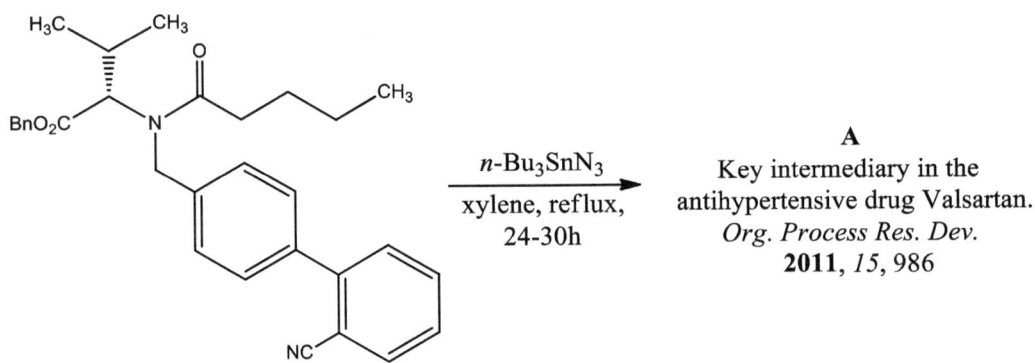

$$\xrightarrow[\substack{\text{xylene, reflux,} \\ 24\text{-}30\text{h}}]{n\text{-Bu}_3\text{SnN}_3}$$

A
Key intermediary in the
antihypertensive drug Valsartan.
Org. Process Res. Dev.
2011, *15*, 986

4)

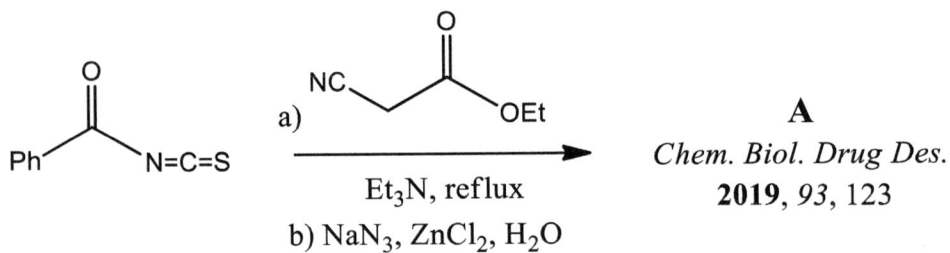

a)

$$\xrightarrow[\substack{\text{Et}_3\text{N, reflux} \\ \text{b) NaN}_3, \text{ZnCl}_2, \text{H}_2\text{O}}]{}$$

A
Chem. Biol. Drug Des.
2019, *93*, 123

Hint: In step (a), a six-membered ring with one nitrogen and one oxygen inside has been formulated.

5)

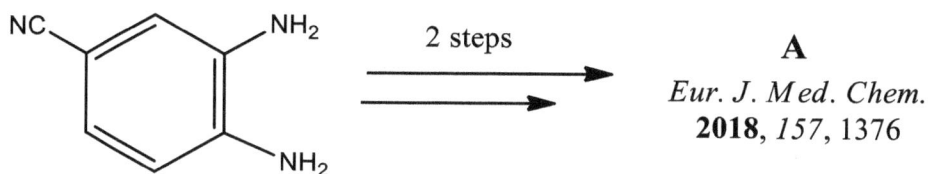

a) 5 M HCl, HCOOH, reflux, 4 h, 94%; b) K$_2$CO$_3$, EtOH, H$_2$NOH.HCl, reflux, 12 h, 50%.

6)

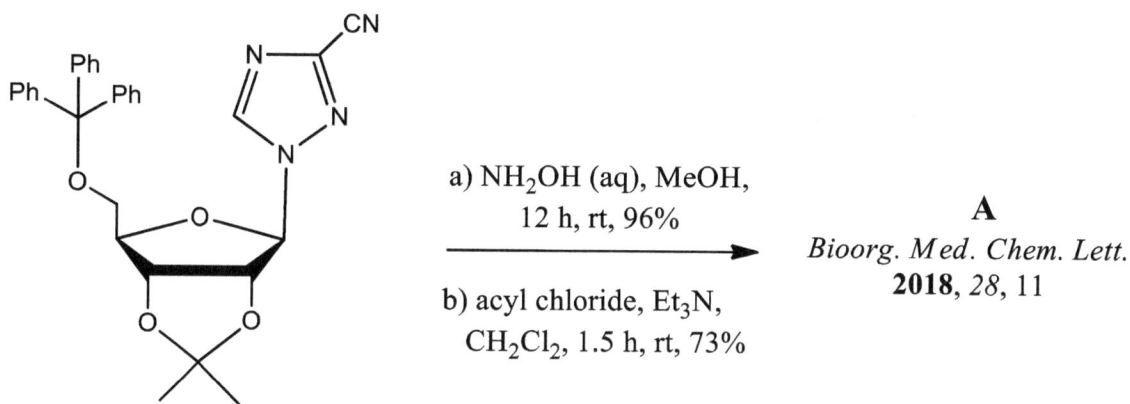

7)

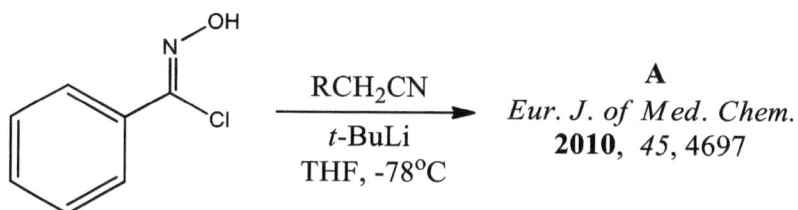

8)

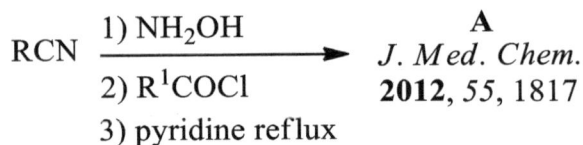

9)

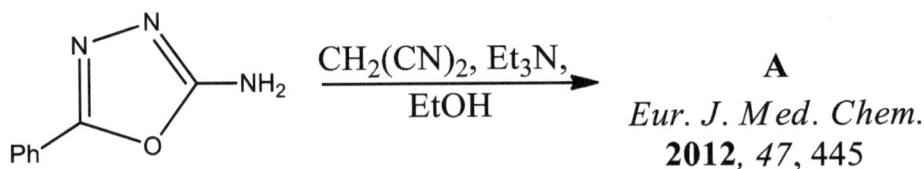

Hint: Formation of a bicyclic compound with the participation of both nitrile groups.

10)

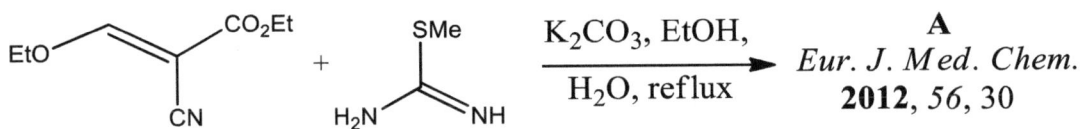

Hint: Formation of the six-membered ring with two nitrogen inside and without the participation of the nitrile group.

11)

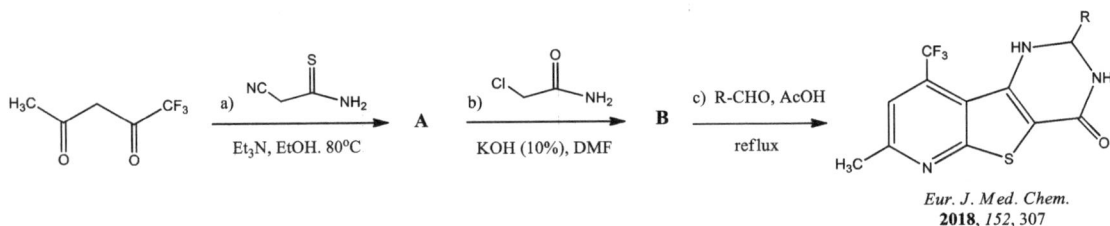

12) What is the structure of the compounds **A** and **B**?

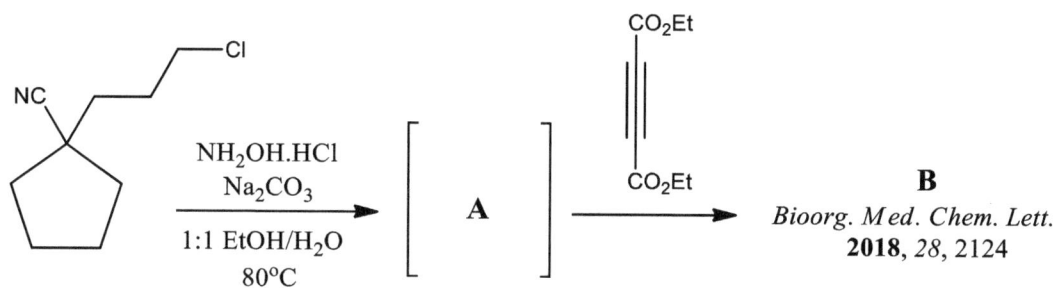

The combination of the functional group's nitrile and hydrazine is advantageous in the formation of heteroaromatic nuclei with important applications in medicinal chemistry. In the exercises below, the structure of the respective heteroaromatics has been identified.

13)

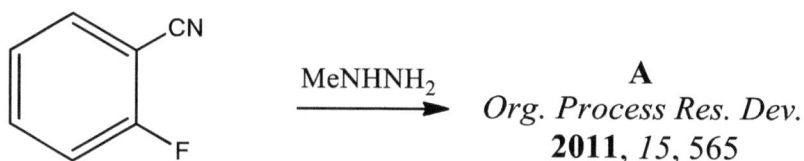

$$\text{MeNHNH}_2 \longrightarrow$$

A
Org. Process Res. Dev.
2011, *15*, 565

14)

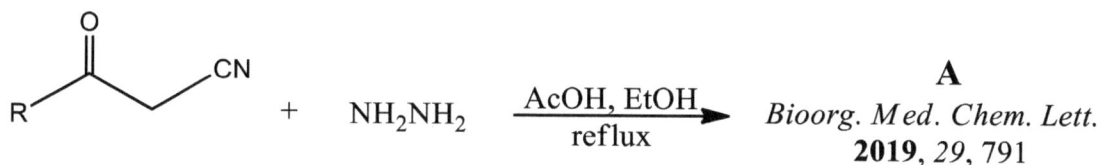

$$+ \quad \text{NH}_2\text{NH}_2 \xrightarrow[\text{reflux}]{\text{AcOH, EtOH}}$$

A
Bioorg. Med. Chem. Lett.
2019, *29*, 791

R = different substituents

15)

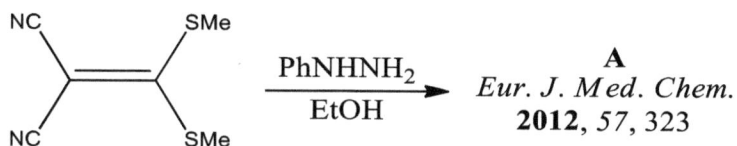

$$\xrightarrow[\text{EtOH}]{\text{PhNHNH}_2}$$

A
Eur. J. Med. Chem.
2012, *57*, 323

16)

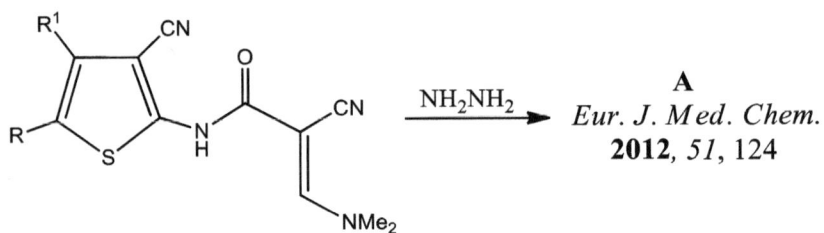

$$\xrightarrow{\text{NH}_2\text{NH}_2}$$

A
Eur. J. Med. Chem.
2012, *51*, 124

17)

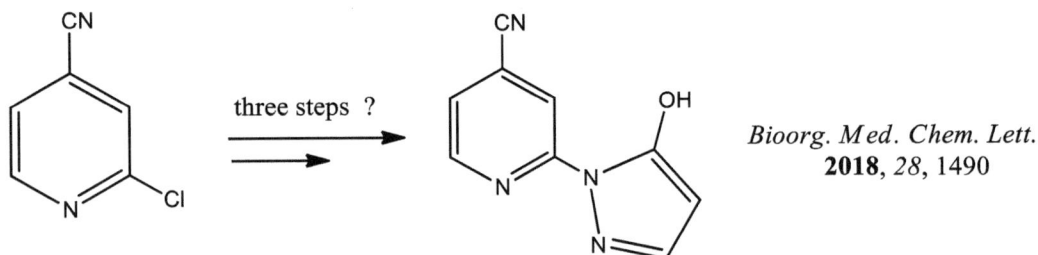

three steps ?

Bioorg. Med. Chem. Lett.
2018, *28*, 1490

Answers

1)

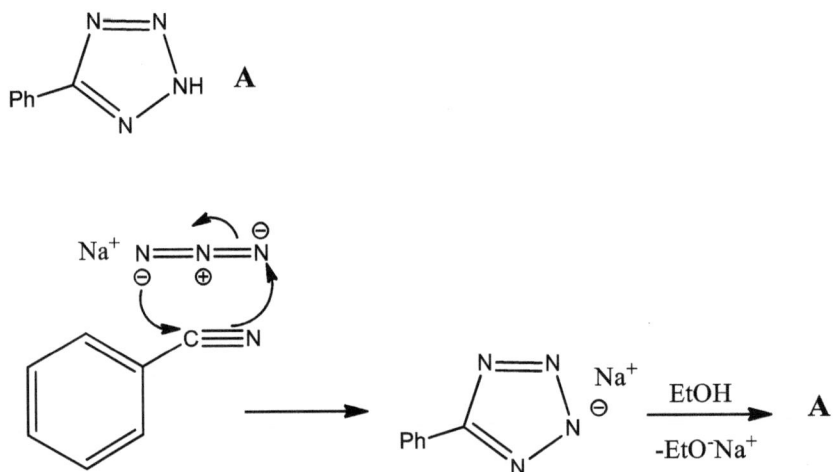

2)

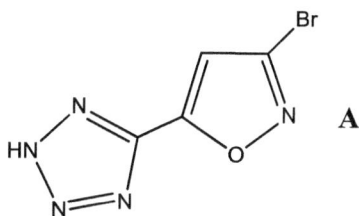

A

3)

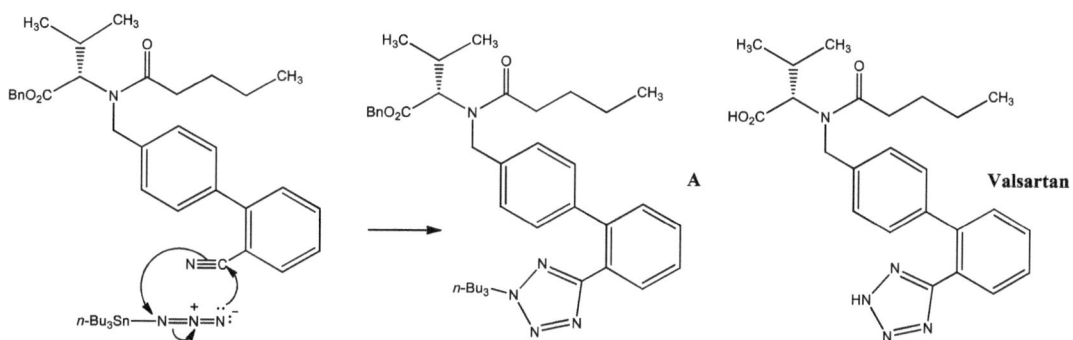

A

Valsartan

4)

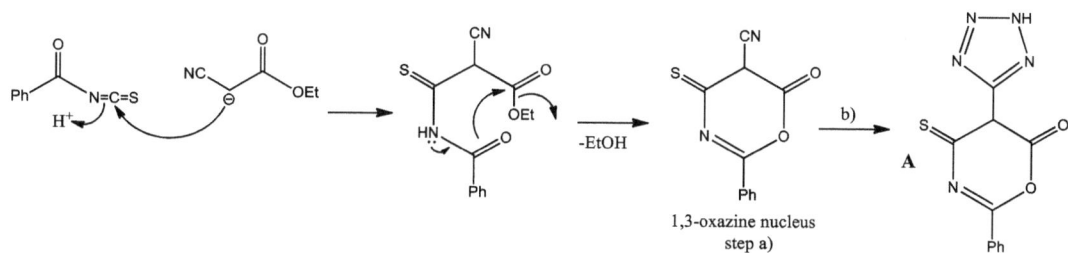

1,3-oxazine nucleus
step a)

A

5)

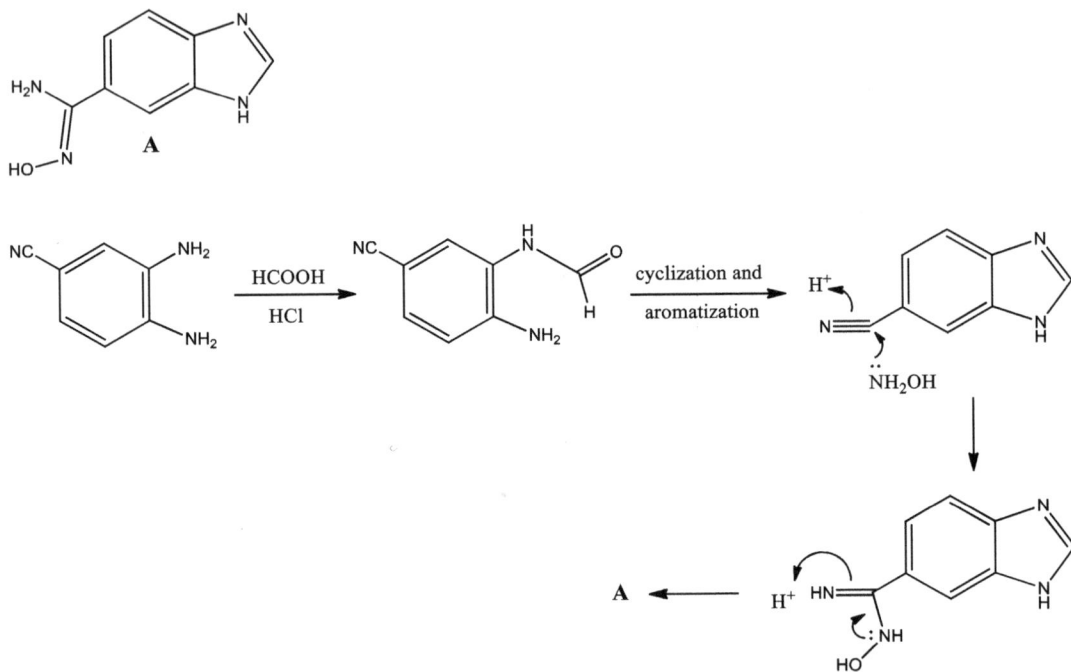

6)

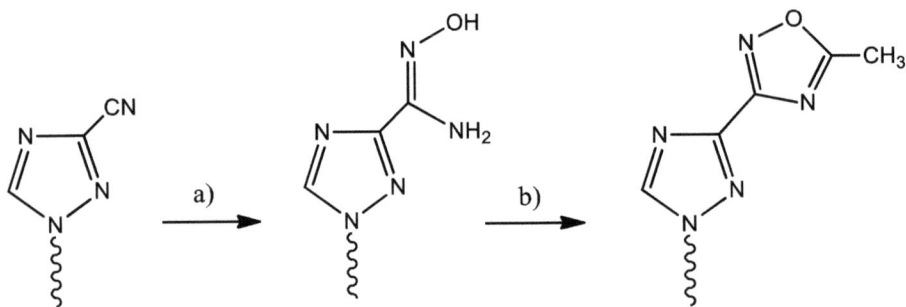

7)

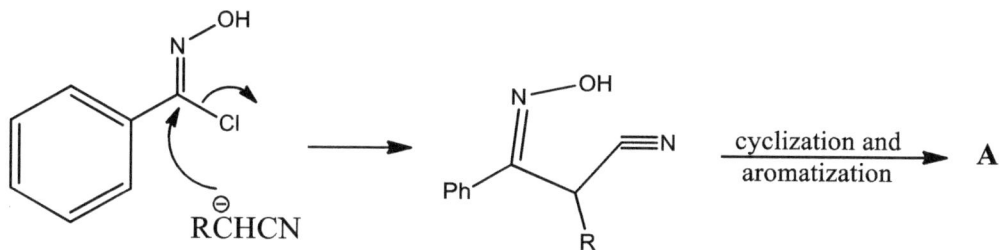

8)

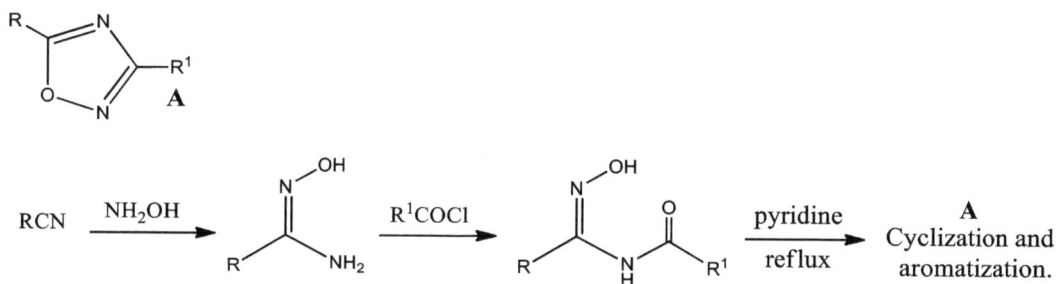

9)

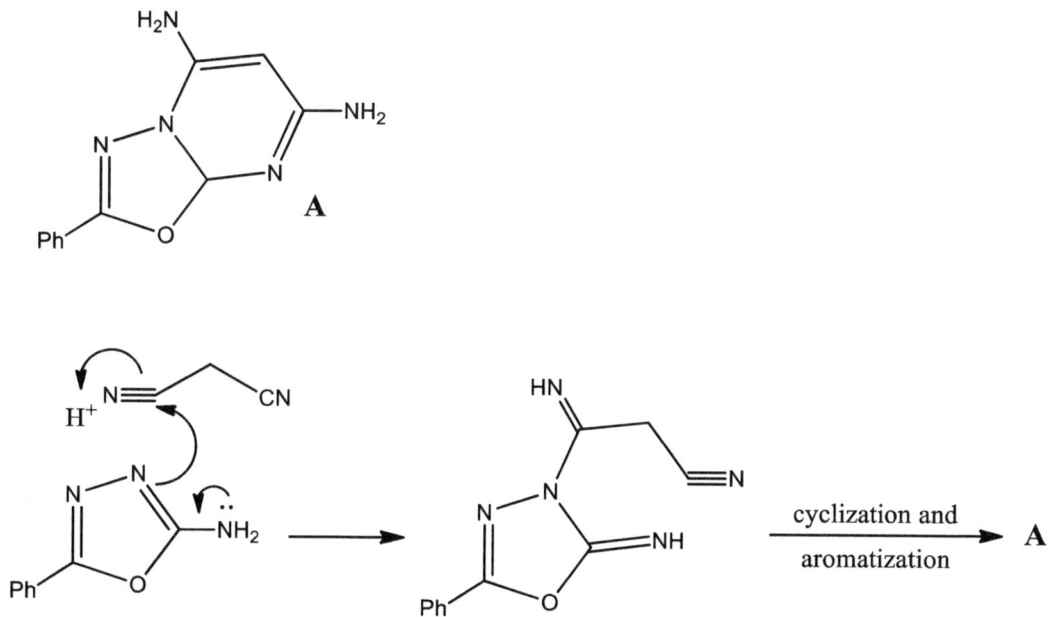

10)

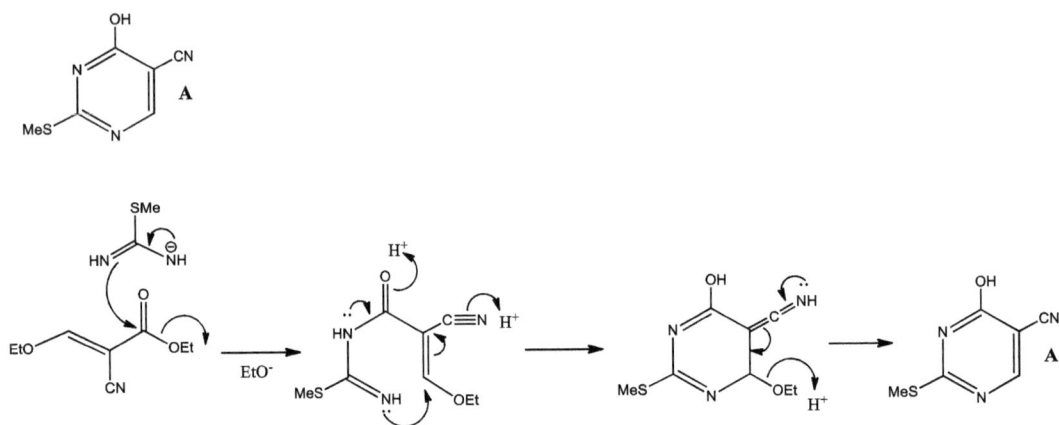

11)

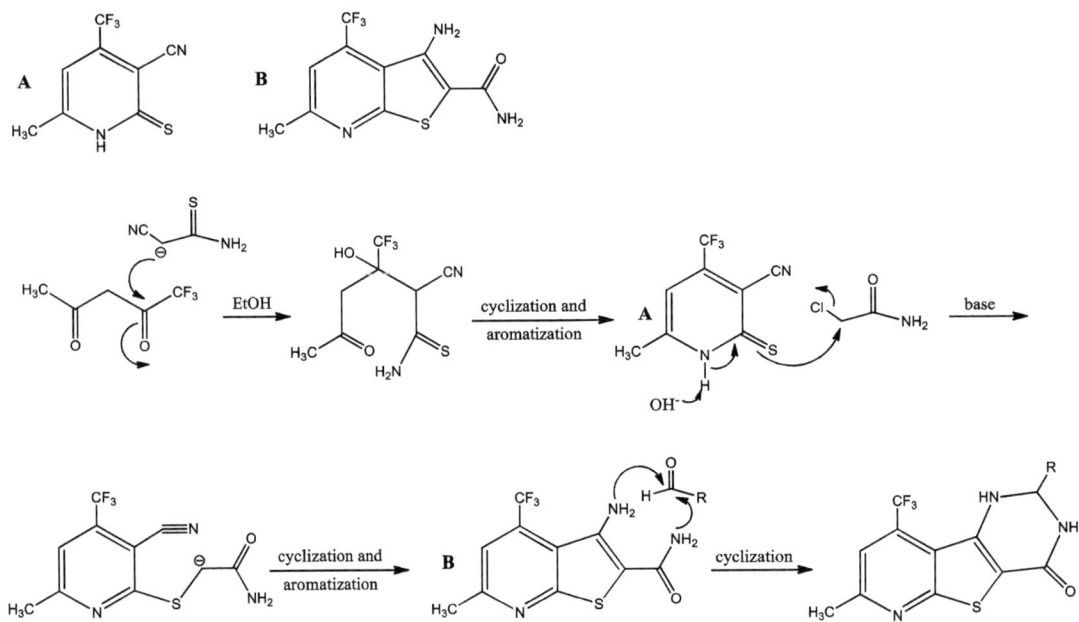

12)

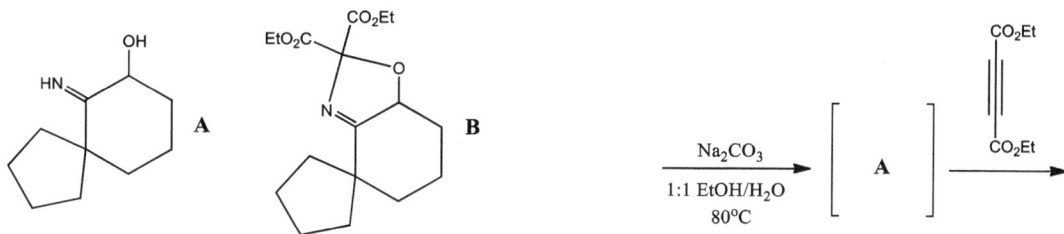

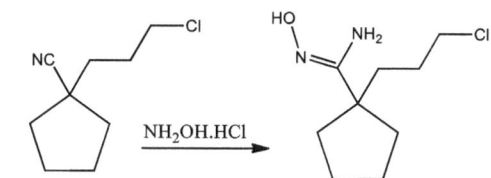

13)

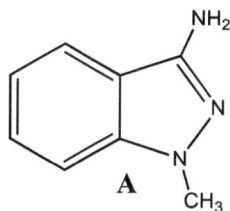

14)

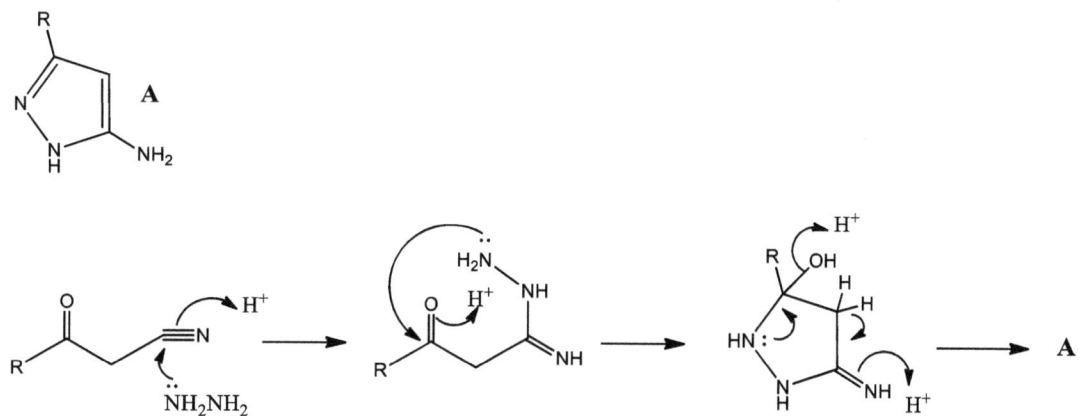

15)

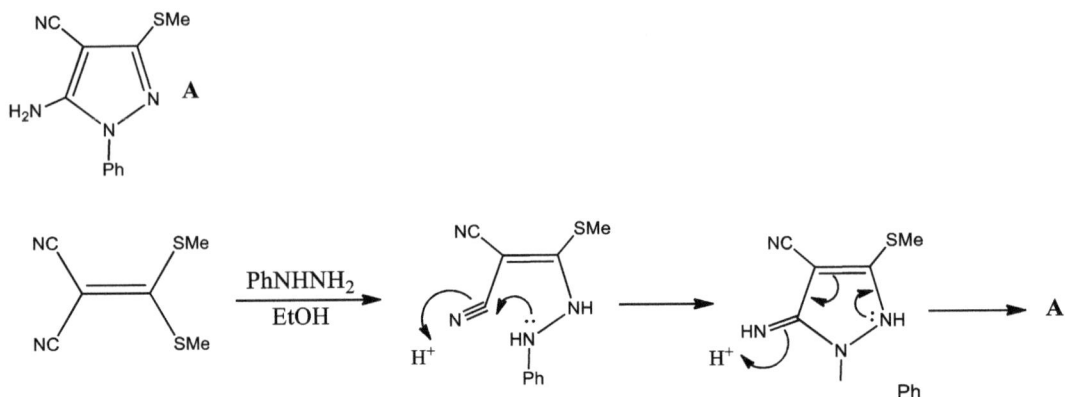

16)

Nucleophilic attack of the hydrazine.

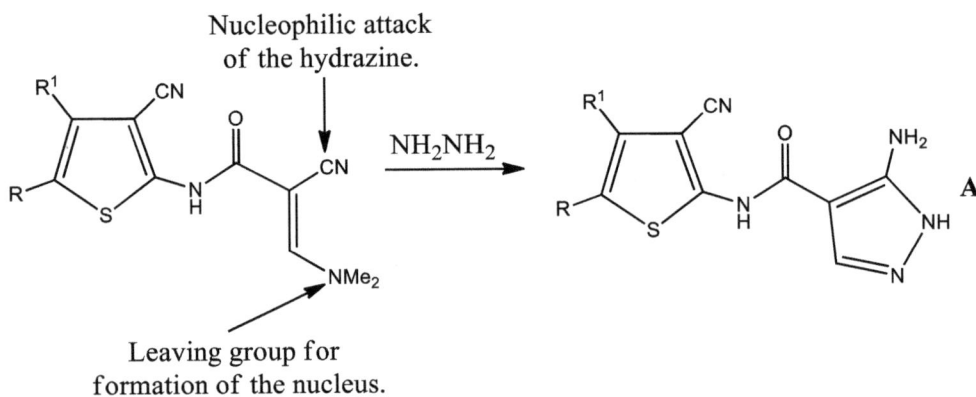

Leaving group for formation of the nucleus.

17)

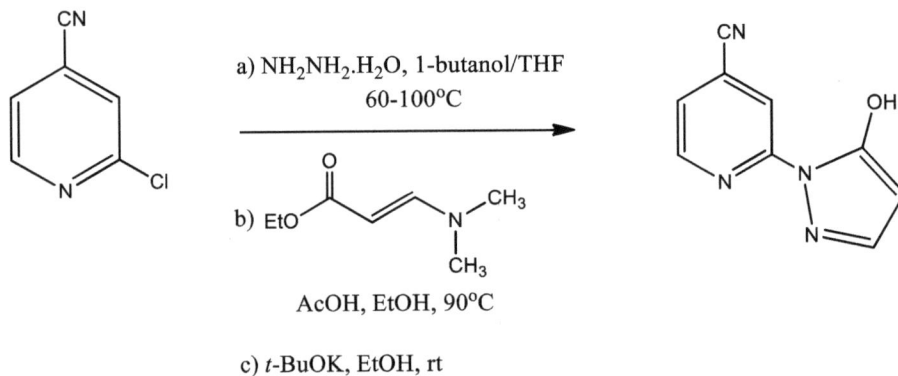

a) $NH_2NH_2 \cdot H_2O$, 1-butanol/THF
60-100°C

b) EtO—...—N(CH$_3$)(CH$_3$)

AcOH, EtOH, 90°C

c) *t*-BuOK, EtOH, rt

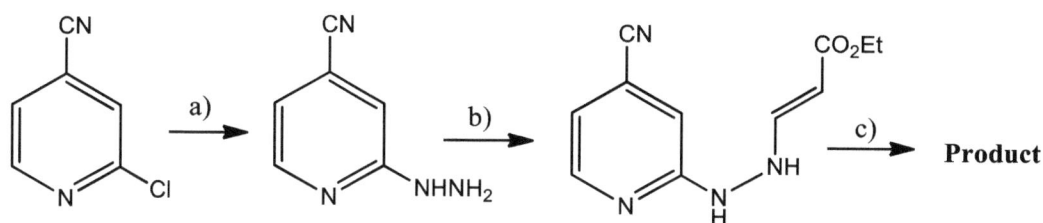

CONCLUSION

The nitrile chemical function has great versatility, being transformed into several other functional groups such as ketones, aldehydes, carboxylic acids, amines, amides, and other functional groups. The use of nitriles is also advantageous in forming heteroaromatic nuclei with essential applications in medicinal chemistry. This class is also found in several drugs in the market for different diseases. Due to its versatility, the exercises in this chapter aim to demonstrate the nitriles as an essential tool in organic synthesis, to the readers. Therefore, it is highly recommended that the reader knows its chemical reactions and applications.

CHAPTER 7

Combinatorial Chemistry

Abstract: Combinatorial chemistry is a methodology developed based on the synthesis of many compounds (hundreds, thousands, or even millions) in a simple chemical process building libraries of these compounds. In medicinal chemistry, an essential factor is the number of different molecules to be produced. For obtaining a leading compound and later a candidate drug, the biological evaluation of thousands of compounds is necessary. Therefore, the exercises below are based on combinatorial chemistry.

Keywords: Bioactive compounds, Combinatorial chemistry, Chemical transformations, Drugs, Exercises, Heteroaromatic compounds, Medicinal chemistry, Molecules, Reaction conditions, Reagents, Retrosynthetic analysis, Substances, Synthesis, Synthons.

INTRODUCTION

In medicinal chemistry, an essential factor is the number of different molecules to be produced. For obtaining a leading compound and later a candidate drug, the biological evaluation of thousands of compounds is necessary, previously provided by using organic syntheses. Combinatorial chemistry is a methodology developed with this purpose and based on the synthesis of a large number of compounds (hundreds, thousands or even millions) in a simple chemical process building libraries of compounds. Scheme **1** shows an example of the concept of combinatorial chemistry by using A-C and 1-9 as reagents using all these reagents together in a single reaction we produced at the same time nine different compounds.

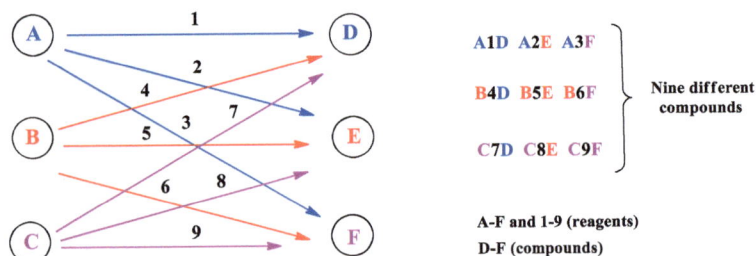

Scheme 1. Example of combinatorial chemistry.

Marcus Vinícius Nora de Souza

Combinatorial chemistry is a methodology developed in the 1960s by Professor Bruce Merrifield (1921-2006) (Scheme **2**). He studied solid-phase of peptides, which introduced the use of insoluble and chemically inert polymers (commonly referred to as resins) as carriers, which are covalently bound to the substrate, in this case, the peptides. The great advantage of solid support synthesis is the simplicity of the purification of the products obtained. The use of a simple filtration followed by successive washing of the solid support with different types of solvents is sufficient for the purification of the product obtained. After the discovery of Merrifield, other researchers developed resins to improve the synthesis in solid support, among which Wang resin is one of the most commonly used. Due to his important work, Merrifield was awarded the Nobel Prize in 1984. In the 1990s, combinatorial chemistry had a major impact on the pharmaceutical industry in the development of new drugs. However, combinatorial chemistry has some disadvantages, such as the production of a mixture of compounds, and it is difficult to isolate and identify the biologically active compound.

Robert Bruce Merrifield

(1921 – 2006)

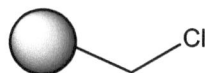

Resin de Merrifield

Resin de Wang

Due to the structural complexity this is usually the representation of the resin.

Nobel Prize 1984

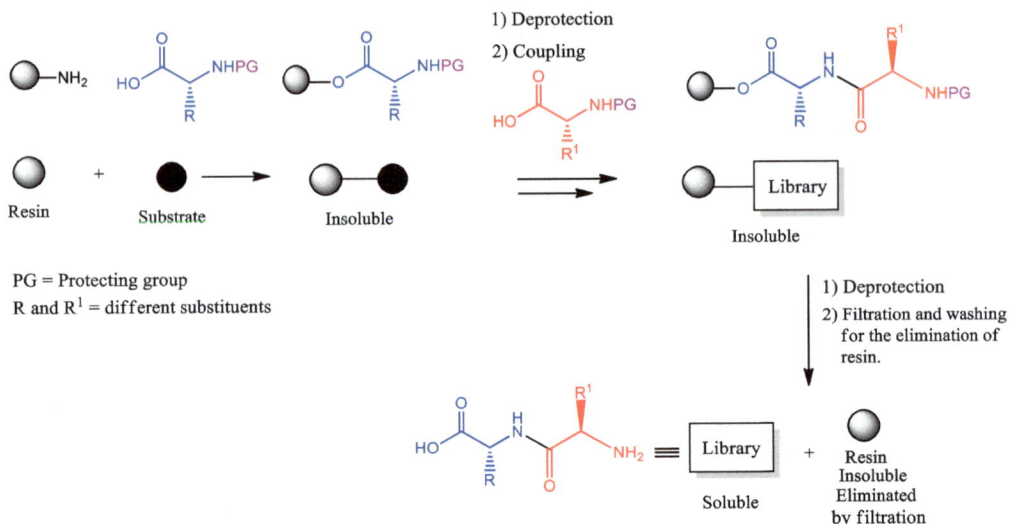

Scheme 2. Principle of solid support synthesis and resins commonly used in solid support synthesis.

Example

Using 4-fluoro-3-nitrobenzoic acid as start material, the product of this reaction was 4-fluoro-3-nitro-benzoate PEGester. The reagents used by the introduction of this resin were *N,N′*-dicyclohexylcarbodiimide (DCC) as the esterification agent and 4-dimethylaminopyridine (DMAP) as a catalyst. The 4-fluoro-3-ni-ro-benzoate PEGester is a key intermediate for the preparation of a library of 1,2,5-trisubstituted benzimidazole derivatives in some steps. What are these steps?

Answer

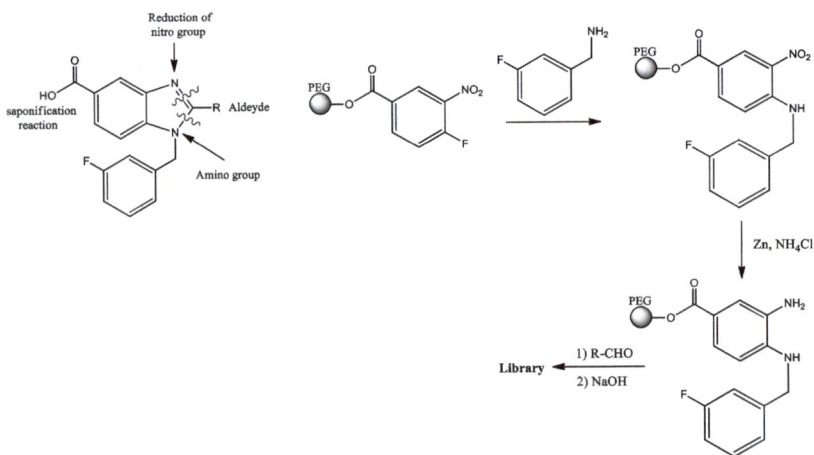

The reaction mechanism of 4-fluoro-3-nitro-benzoate PEGester is summarized below. If the reader wants to know more about this reaction, it is recommended to consult the recommended bibliography [1, 2]. DCC (*N,N*'-dicyclohexyl carbodiimide) and DMAP (*N, N*-dimethylaminopyridine) are used as a catalyst in standard conditions for the formation of amides and esters employed in peptide chemistry. However, these conditions present in some cases drawbacks, which is due to the formation of DCU (*N,N*'-dicyclohexylurea), a problematic by-product challenging to eliminate. The use of DMAP as a catalyst can also present problems. Due to its basicity, DMAP can in certain cases racemize stereogenic centers. In the attempt to solve these problems, new reagents such as EDC (1-ethyl-3-(3-dimethylaminopropyl) carbodiimide hydrochloride have been developed, which produce as a byproduct water-soluble urea. This urea is much easier to eliminate from the reaction than DCU. The use of HOBt (1-hydroxybenzotriazole) as a catalyst instead of DMAP is preferred because it is less basic and nucleophilic [1, 2].

MECHANISM OF REACTION

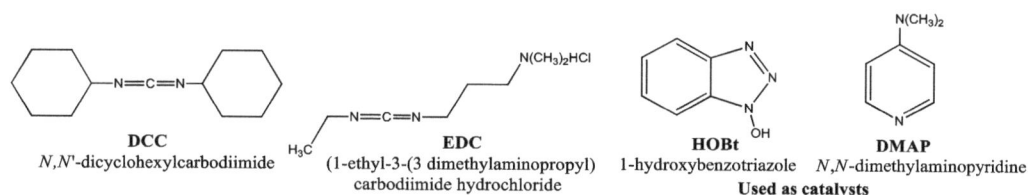

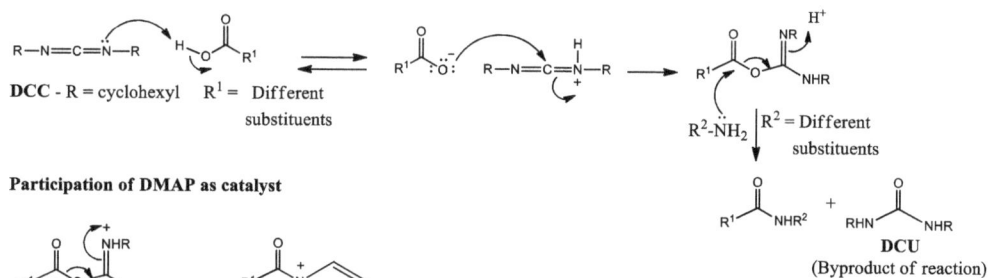

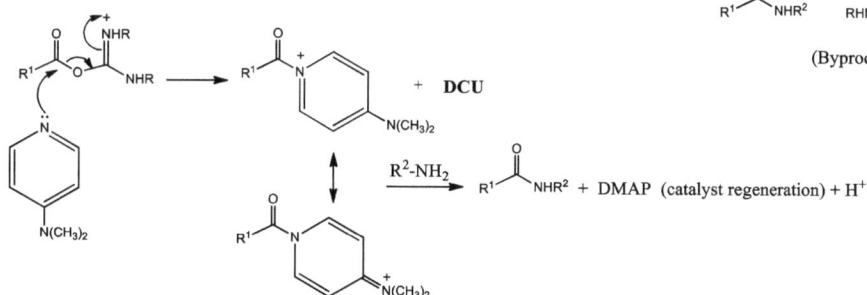

EXERCISES

1) Benzodiazepines are a known class of psychoactive drugs, and, this work built a library of this class by using solid-phase synthesis. Based on its combinatorial synthesis, what is your basic structure?

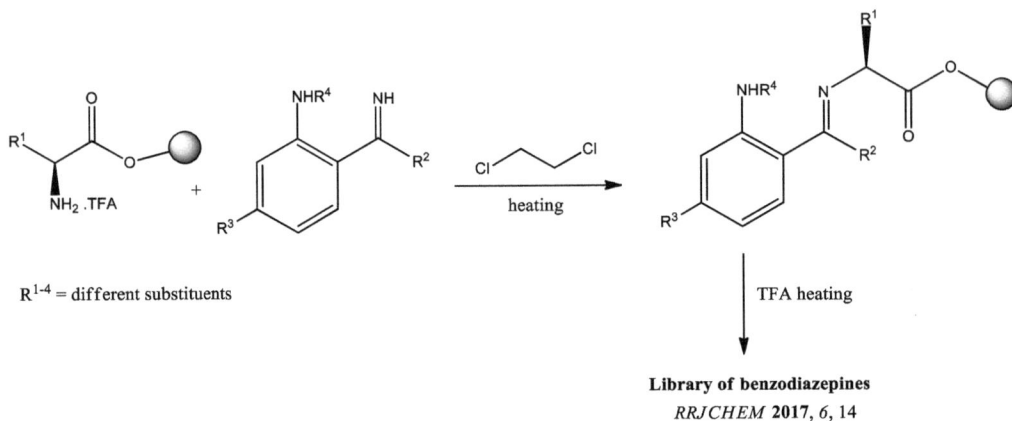

R^{1-4} = different substituents

Library of benzodiazepines
RRJCHEM **2017**, *6*, 14

2) What is the structure of the drug Prazosin?

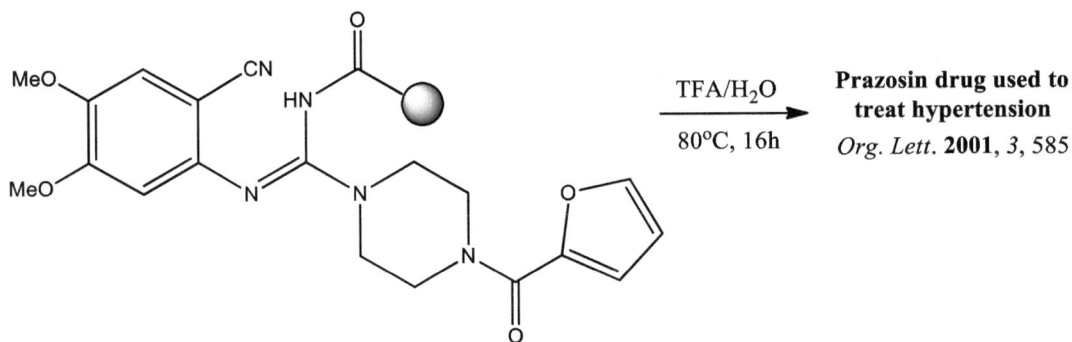

TFA/H$_2$O

80°C, 16h

Prazosin drug used to treat hypertension

Org. Lett. **2001**, *3*, 585

3) What is the structure of this heteroaromatic nucleus?

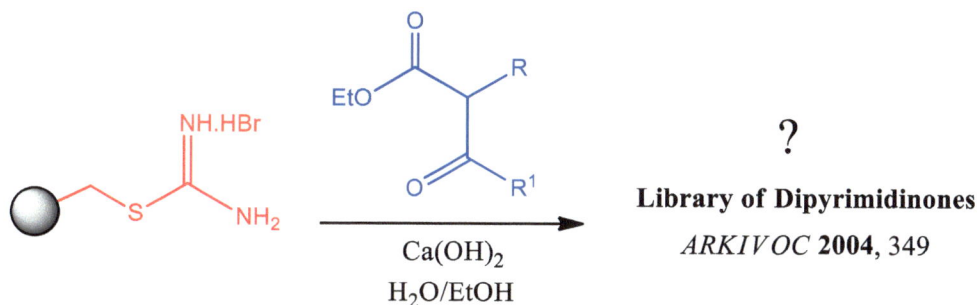

?

Library of Dipyrimidinones

ARKIVOC **2004**, 349

4) Hydantoins are a class of useful building blocks that have a wide range of applications, such as in medicinal chemistry. How would you synthesize in three steps a library of this class?

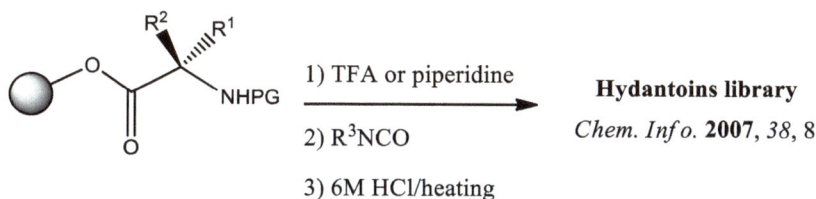

1) TFA or piperidine

2) R^{3}NCO

3) 6M HCl/heating

Hydantoins library

Chem. Info. **2007**, *38*, 8

R^{1,2} = different substituents

PG = Protecting group

5) Using solid-phase synthesis, how to produce a library of 4,6-diaryl-3-4-dihydropyrimidine derivatives?

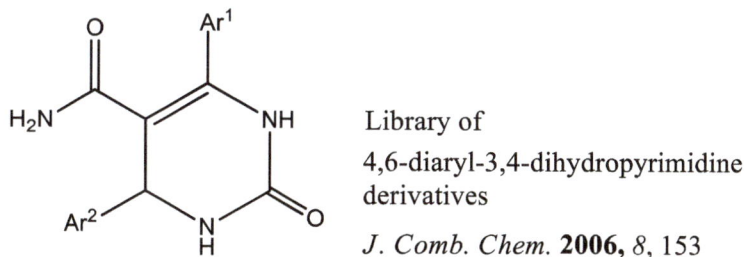

Library of 4,6-diaryl-3,4-dihydropyrimidine derivatives

J. Comb. Chem. **2006**, *8*, 153

6) A useful methodology for the synthesis of the pyrimidine nucleus is the reaction between alpha, beta-unsaturated ketone, and guanidine. With this information in hand, can resin Merrifield and *p*-hydroxybenzaldehyde produce a library of this class?

Trisubstituted Pyrimidines

ACS Comb. Sci. **2019**, *21*, 35

Bioorg. Med. Chem. Lett. **2005**, *15*, 4923

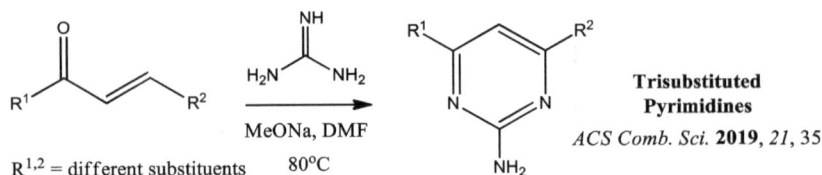

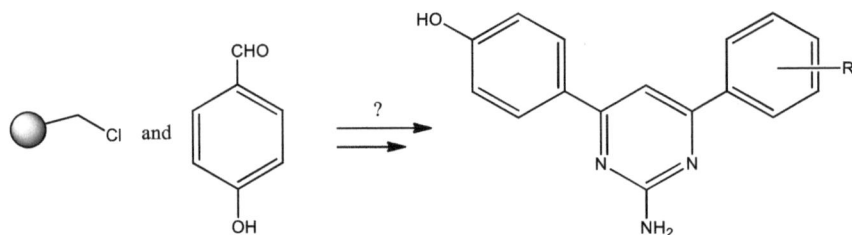

It is important to mention that it is also possible to build a library of compounds without using a resin. The exercises below are good examples of this strategy.

7) Grigorenko and co-workers developed a multigram scale of (chlorosulfonyl) benzenesulfonyl fluorides, successfully applied in the synthesis of a library of sulfonamide derivatives. This library was tested as serine protease inhibitors using trypsin as a model enzyme and the researchers found the lead compound **2**. Using compound **1** how would you synthesize this critical building block?

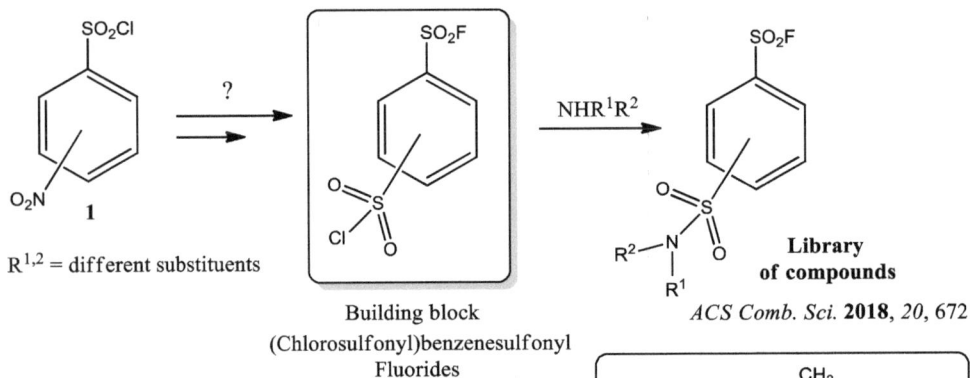

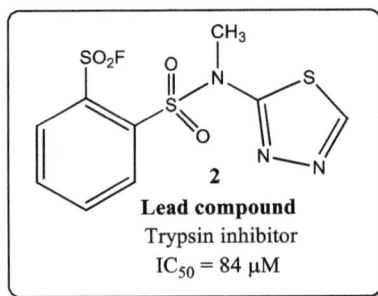

Building block
(Chlorosulfonyl)benzenesulfonyl
Fluorides

ACS Comb. Sci. **2018**, *20*, 672

Library of compounds

2

Lead compound
Trypsin inhibitor
IC_{50} = 84 μM

8) What is the structure of this class of compound?

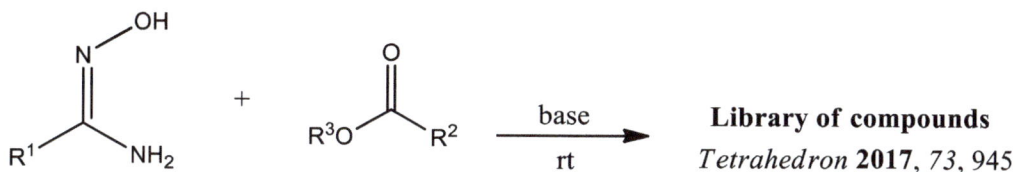

R^{1-3} = different substituents

9) Below we have the synthesis of a library of imidazo[1,2.a]pyridines 1-tetrazolylimidazo[1,5-a]pyridine. Propose the mechanism of this reaction.

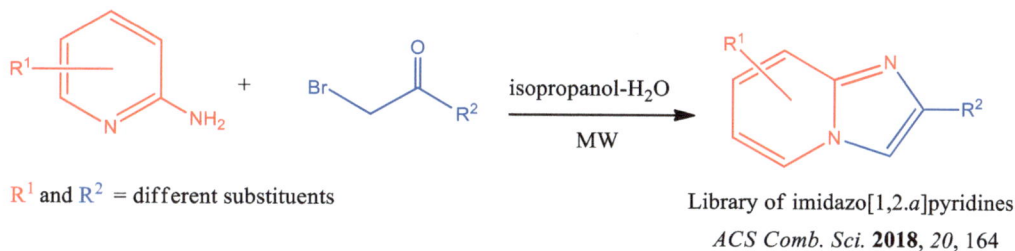

R^1 and R^2 = different substituents

Library of imidazo[1,2.*a*]pyridines
ACS Comb. Sci. **2018**, *20*, 164

Answers

1)

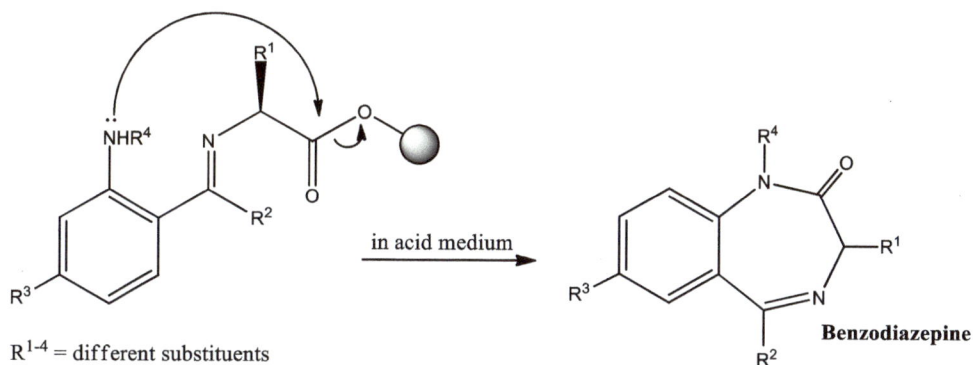

R^{1-4} = different substituents

in acid medium

Benzodiazepine

2)

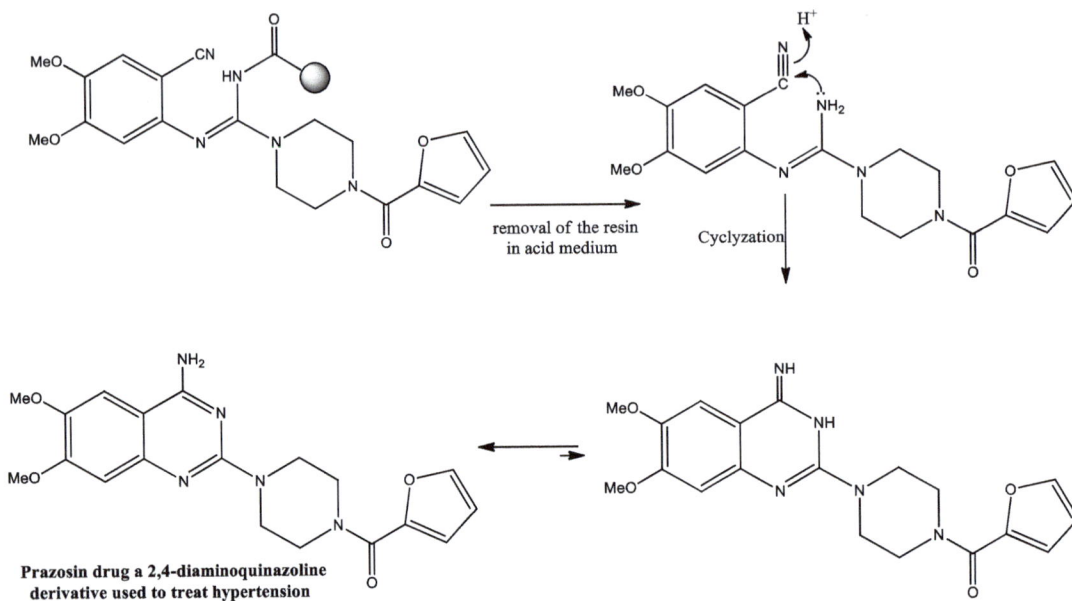

Prazosin drug a 2,4-diaminoquinazoline
derivative used to treat hypertension

3)

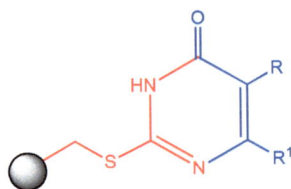

4)

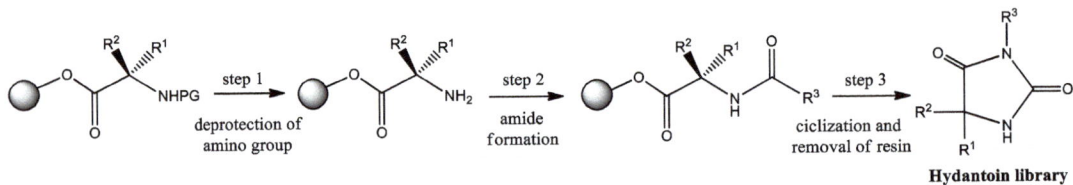

Hydantoin library

5)

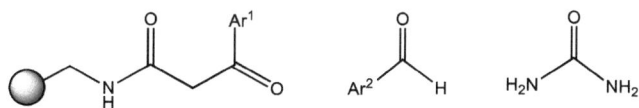

6)

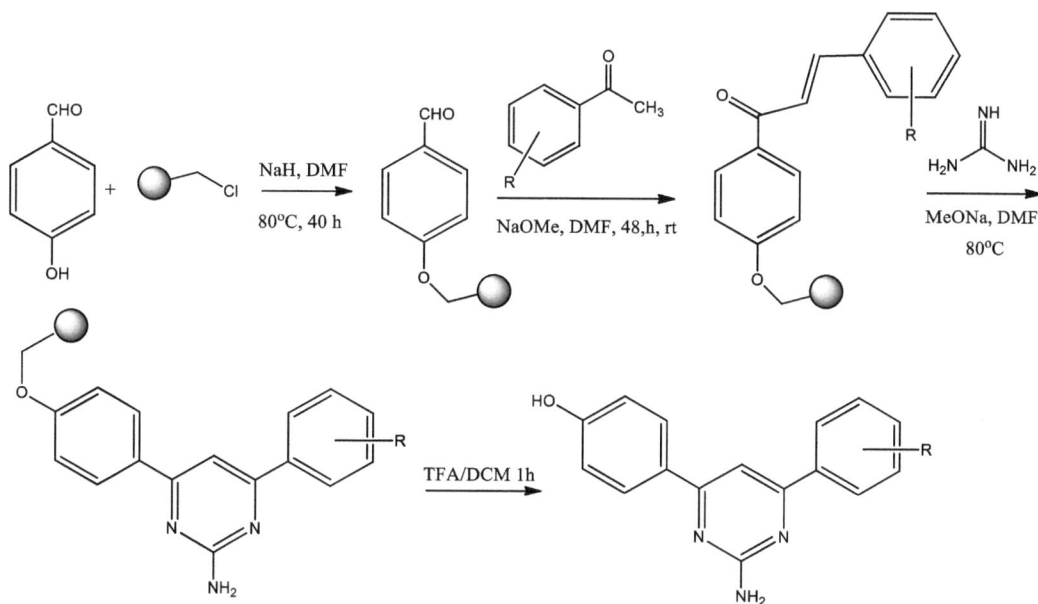

7)

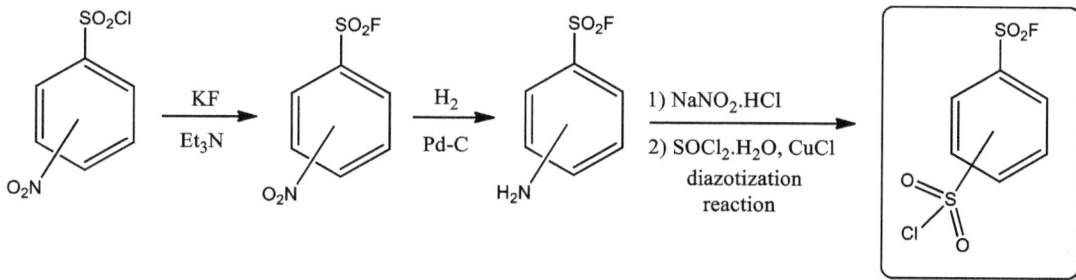

8)

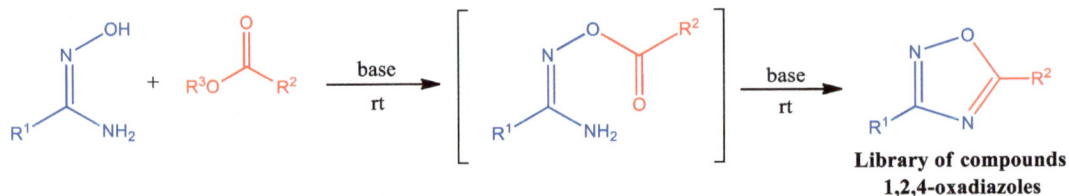

Library of compounds
1,2,4-oxadiazoles

9)

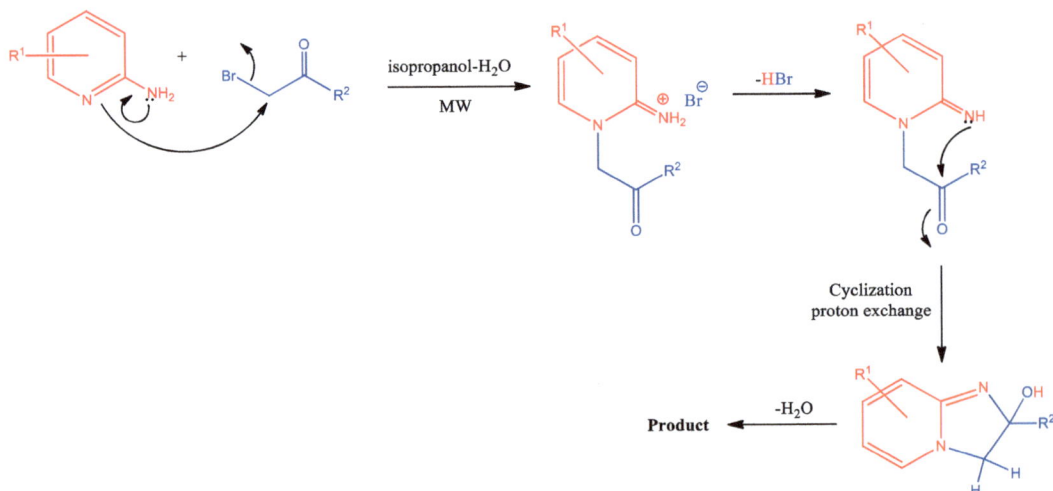

Cyclization
proton exchange

Product

CONCLUSION

In medicinal chemistry, an essential factor is the number of different molecules produced and then tested. Considering that, combinatorial chemistry is a methodology developed based on the synthesis of many compounds (hundreds, thousands, or even millions) in a simple chemical process building libraries of compounds. The exercises based on this concept aim to inform the reader about the importance of this theme for organic synthesis and medicinal chemistry.

REFERENCES

[1]   Valeur, E.; Bradley, M. Amide bond formation: beyond the myth of coupling reagents. *Chem. Soc. Rev.,* **2009**, *38*(2), 606-631.
[http://dx.doi.org/10.1039/B701677H] [PMID: 19169468]

[2]   Joullié, M.M.; Lassen, K.M. Evolution of amide bond formation. *ARKIVOC,* **2010**, *8*, 189-250.
[http://dx.doi.org/10.3998/ark.5550190.0011.816]

CHAPTER 8

Multicomponent Reaction (MCR)

Abstract: Multicomponent reaction (MCR) is a type of reaction based on a mixture of three or more compounds or reagents leading only to one or a major molecule being able to form complex molecules with a high degree of control of the stereogenic centers in only one step. This methodology is also a useful tool in medicinal chemistry to produce a diversity of compounds for biological evaluation. Due to its importance, the exercises below focus on this technique.

Keywords: Bioactive compounds, Chemical transformations, Combinatorial chemistry, Drugs, Exercises, Heteroaromatic compounds, Medicinal chemistry, Molecules, Multicomponent reaction (MCR), Reaction conditions, Reagents, Retrosynthetic analysis, Substances, Synthesis, Synthons.

INTRODUCTION

This methodology is a type of reaction based on a mixture of three or more compounds or reagents leading only to one or a major molecule being able to form complex molecules with a high degree of control of the stereogenic centers in only one step. The first reaction described in the literature based on the principles of the MCR was the Streck reaction. This reaction is the synthesis of α-amino acids by using aldehydes, HCN, and NH_3 as reagents described in 1850 by Adolph Strecker (1822-1871) (Scheme **1**). After Strecker´s work, other scientists developed other types of reactions, such as two reactions based on the utilization of the functional group isocyanide. The first one is the Passerini reaction (1921) (Scheme **2**), which is the synthesis of α-acyloxy carboxamides in a one-pot reaction [1]. The reagents of this reaction are aldehyde or ketone, carboxylic acid, and isocyanides in a 3-component MRC. The second is the Ugi reaction (1959) (Scheme **2**), producing α-aminoacyl amides by using primary amines, aldehydes, carboxylic acids, and isocyanides as starting materials in a 4-component MCR [2]. It is essential to mention that this synthesis and its variations are the most known and used MCRs. The proposed mechanisms of these reactions are shown in Schemes **3** and **4**.

Marcus Vinícius Nora de Souza

Adolph Strecker (1822 – 1871)

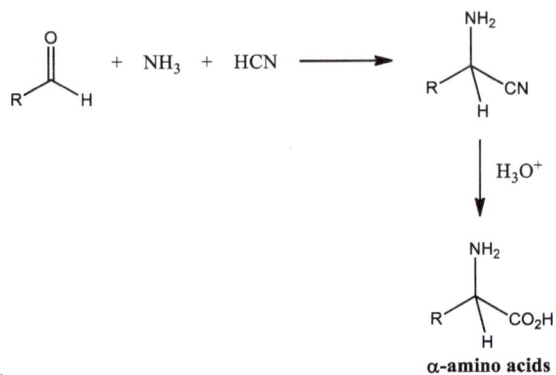

Scheme 1. Synthesis of α-amino acids by using aldehydes, HCN, and NH_3 as reagents, known as Strecker reaction.
Source of the picture: Pedersen, B. Physical Science in Oslo, Phys. Perspect., 2011, 13(2), 215-238.

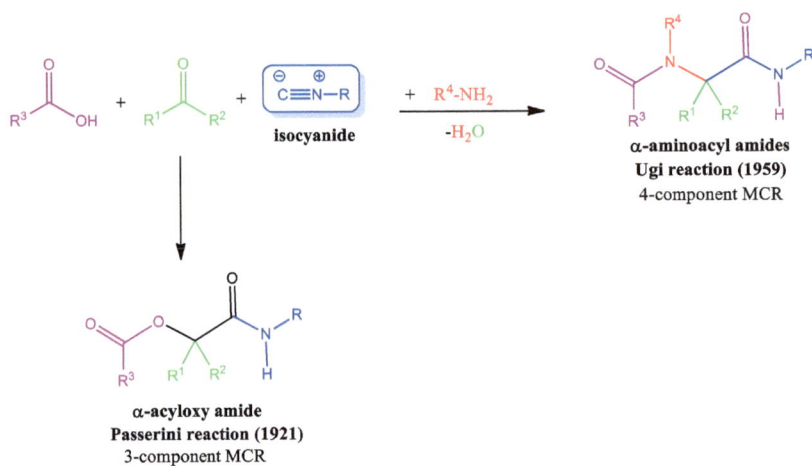

Scheme 2. MCRs developed by Passerini and Ugi.

Passerini reaction

Ionic mechanism base on polar solvents
such as water and methanol

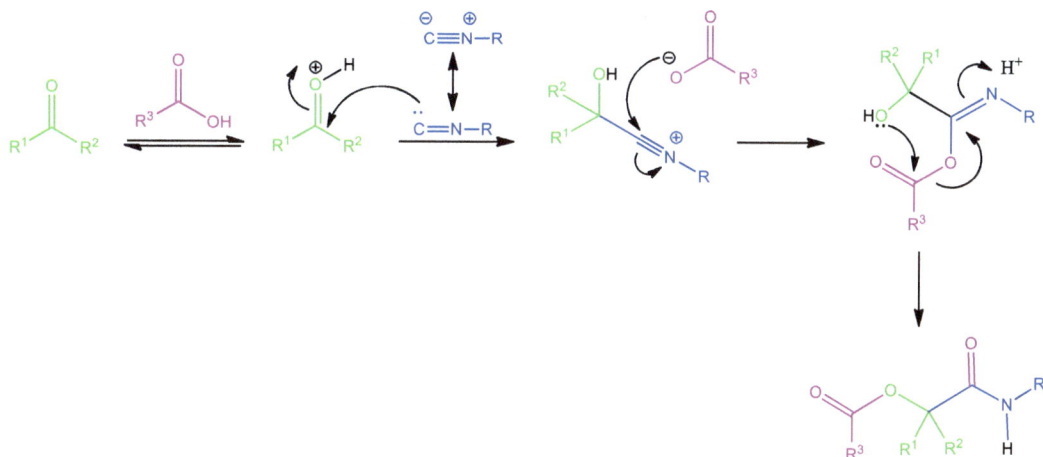

Passerini reaction

Concerted mechanism base on non-polar
solvents and high concentration.

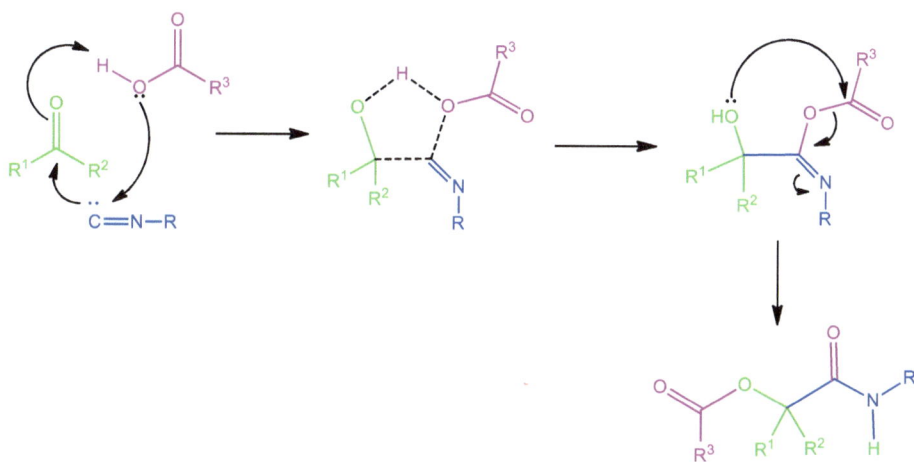

Scheme 3. Mechanism of Passerini reaction.

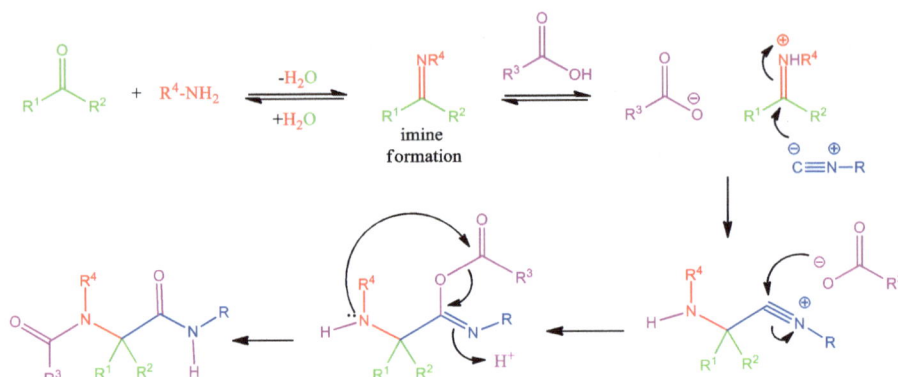

Scheme 4. Mechanism of Ugi reaction.

Another reaction considered as a three multicomponent reaction is the well-known reaction of Mannich [3]. This reaction is the synthesis of β-amine carbonyl substances using ketone, formaldehyde, and amine (Scheme **5**). The mechanism of this reaction is the nucleophilic addition of the amine in question on formaldehyde. After that, the elimination of the hydroxyl group produces the imine functional group (Scheme **5**). This imine is known as Schiff base or azomethine, due to German researcher Hugo Schiff (1834-1915), who discovered this class of substances. The next step consists of condensation between the previously obtained Schiff base and the enolate produced by the respective carbonyl in the presence of a base, which furnished the respective β-aminated carbonyl substance 1. The use of retrosynthetic analysis is also a valuable tool to identify this type of reaction (Scheme **5**).

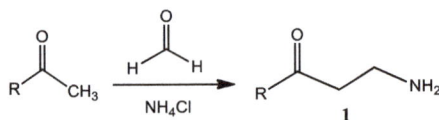

Carl Mannich (1877-1947)

Source of the picture : Link, A. Centennial Anniversary of Mannich's Report on the Formation of β-Amino-Ketones in the Archiv der Pharmazie. Arch. Pharm., **2017**, 350(7), e1700152.

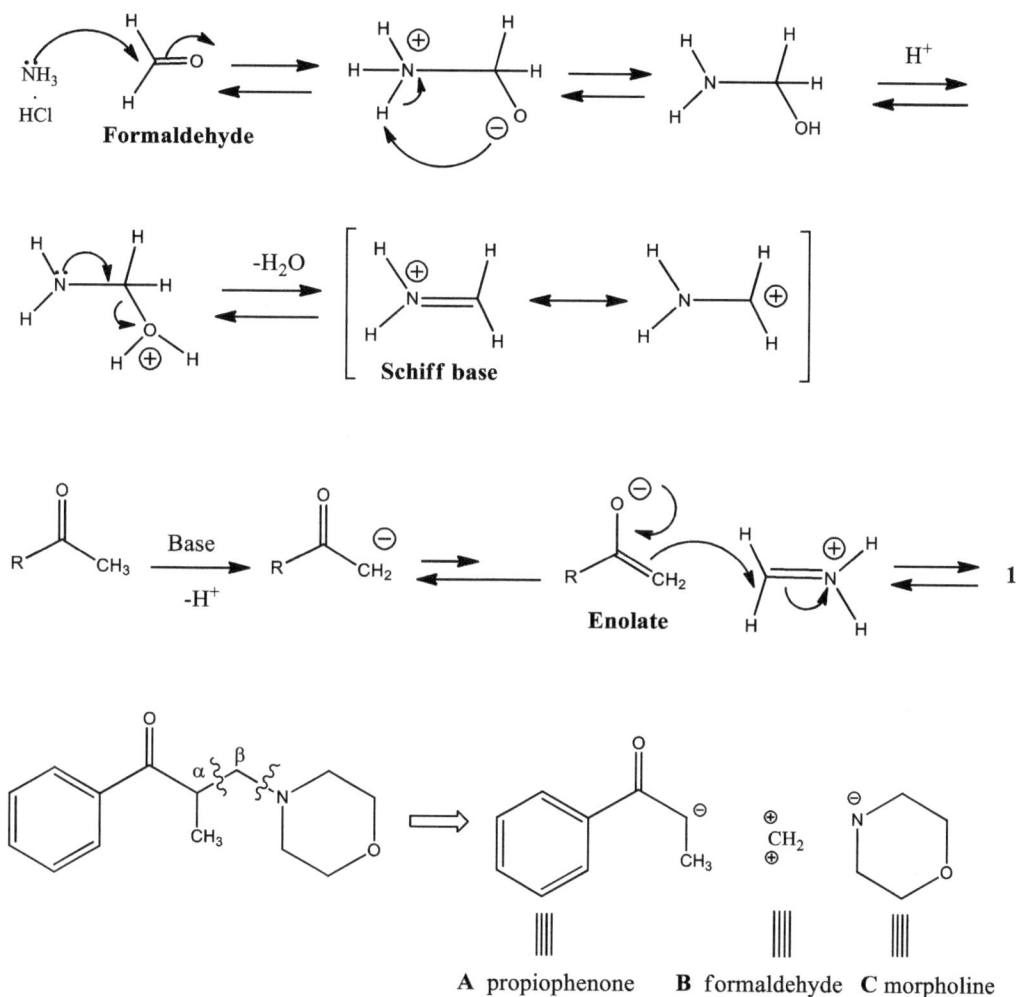

Scheme 5. Mechanism of Mannich reaction and an example of how to identify this type of reaction by using retrosynthetic analysis.

Example

Identify the reagents used in the multicomponent reaction below for the formation of the respective product.

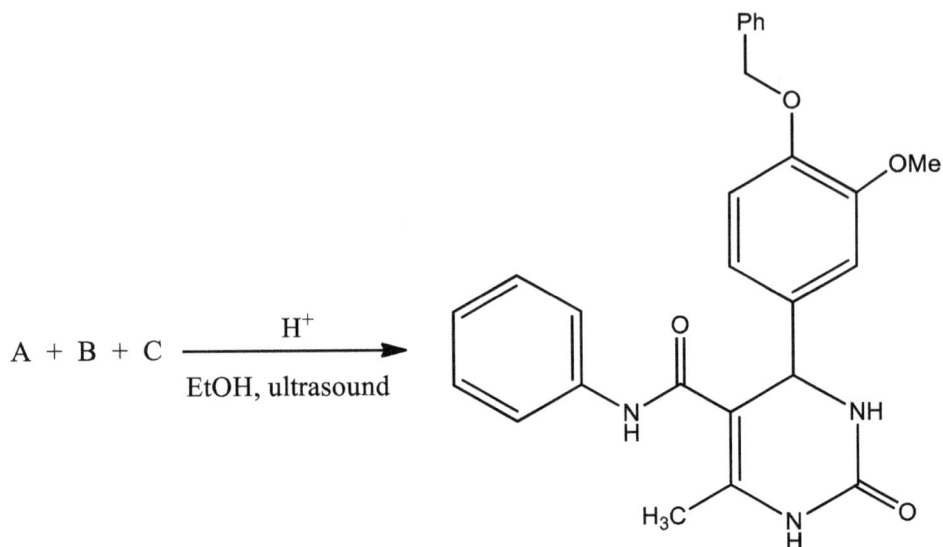

$$A + B + C \xrightarrow[\text{EtOH, ultrasound}]{H^+}$$

Chemistry & Biology Interface **2016**, *6*, 333

Answer

This reaction is known as Biginelli MCR for the preparation of 3,4-dihydropyrimidin-2(1*H*)-ones derivatives. We can identify three functional groups present in the respective substances, urea, diketone, and aldehyde, which react to obtain the target molecule.

Important Note

When we analyze MCR reactions, there are some functional groups and reagents that should always be taken into account. One of the most important is the aldehyde, which generally is present in a vast majority of MCR. Its identification is not so simple but usually appears in reactions of condensation and formation of rings. Other functional groups are amines, ketones, diketones, alkynes, carboxylic acids, and isocyanide. This last one is customarily used for the synthesis of the amides (Passerini reaction). Malononitrile (NC(CH$_2$)CN) is a reagent that also is common in MCR.

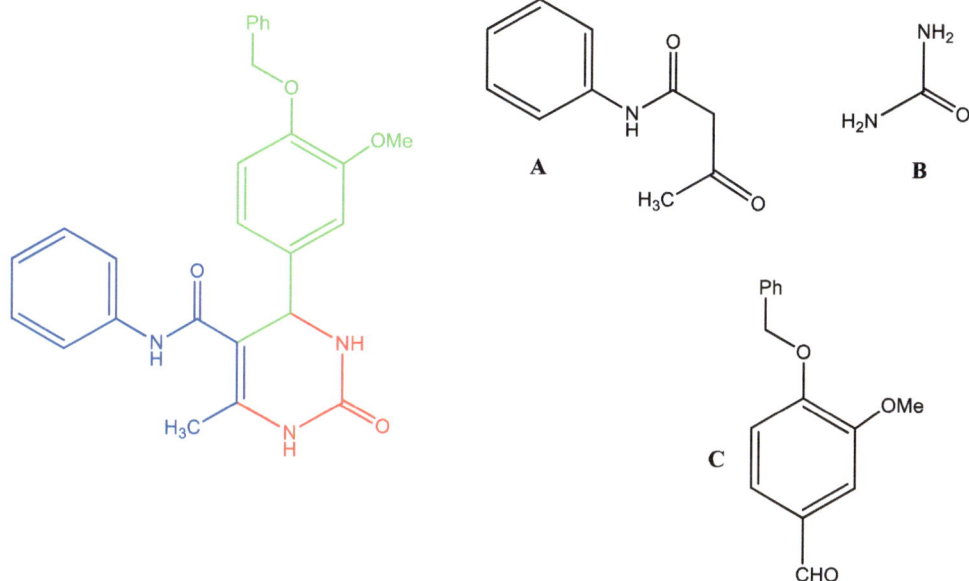

In the exercises below, the reagents and substances identified in multicomponent reactions are used in the synthesis of their respective products. In some activities, the product has colors that will give tips for the resolution of the exercise.

EXERCISES

1)

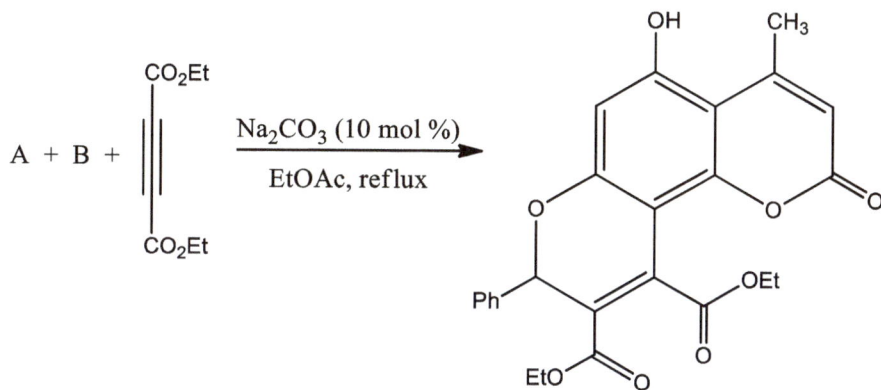

Tetrahedron Lett. **2019**, *60*, 557

2)

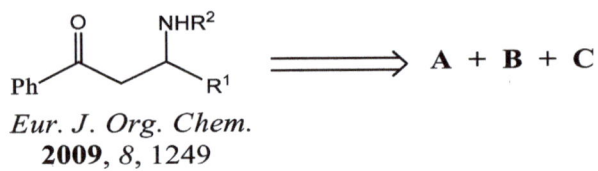

Ph, O, NHR^2, R^1

$$\Longrightarrow \quad \textbf{A} + \textbf{B} + \textbf{C}$$

Eur. J. Org. Chem.
2009, *8*, 1249

3)

$$2\textbf{A} + \textbf{B} + \textbf{C} \xrightarrow[\text{24h, MeOH, reflux}]{\text{H}_2\text{NSO}_3\text{H (30 mol\%)}}$$

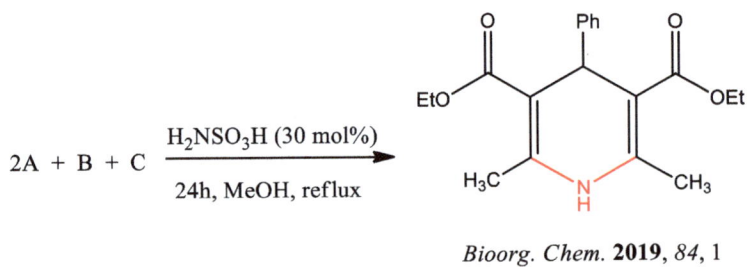

Bioorg. Chem. **2019**, *84*, 1

4)

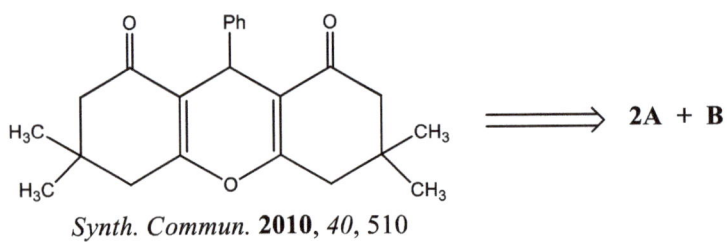

$$\Longrightarrow \quad 2\textbf{A} + \textbf{B}$$

Synth. Commun. **2010**, *40*, 510

5)

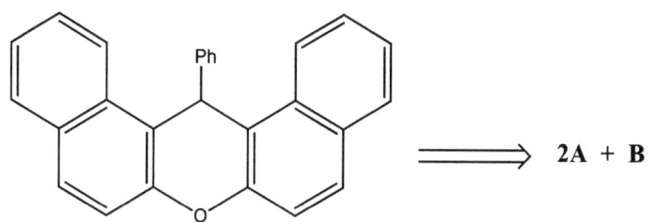

Synth. Commun. **2009**, *39*, 4328

$$\Longrightarrow \quad 2A + B$$

6)

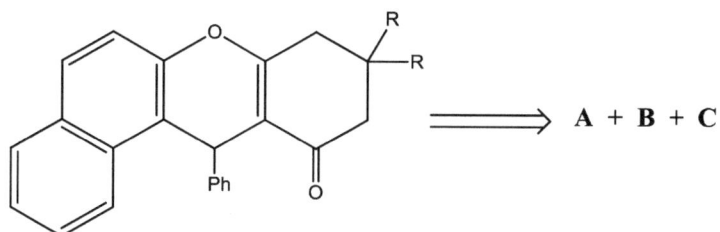

Synth. Commun. **2009**, *39*, 2101

$$\Longrightarrow \quad A + B + C$$

7)

$$A + B + C \longrightarrow$$

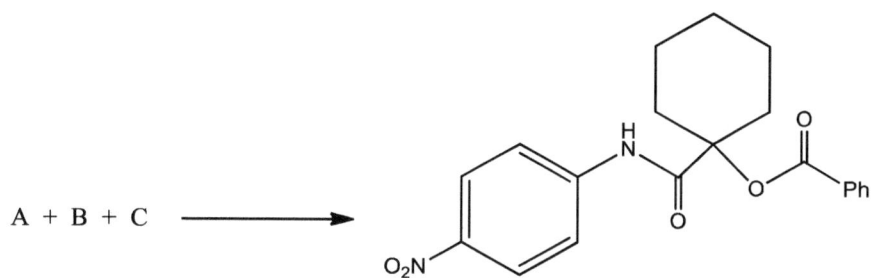

Eur. J. .Med. Chem. **2019**, *168*, 340

8)

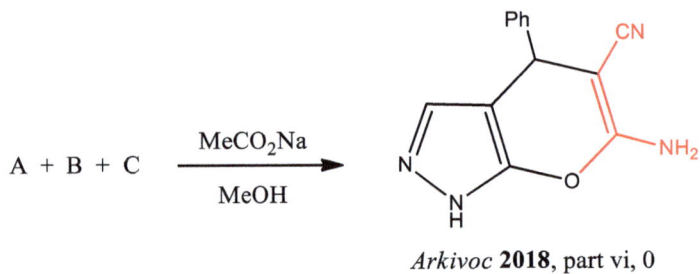

$$A + B + C \xrightarrow[\text{MeOH}]{\text{MeCO}_2\text{Na}}$$

Arkivoc **2018**, part vi, 0

9)

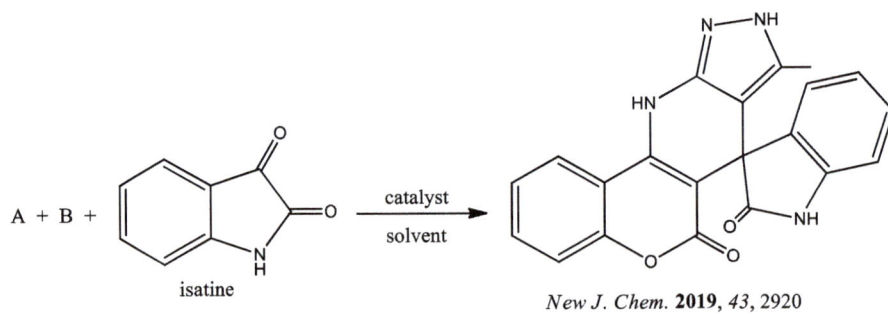

$$A + B + \text{isatine} \xrightarrow[\text{solvent}]{\text{catalyst}}$$

isatine

New J. Chem. **2019**, *43*, 2920

10)

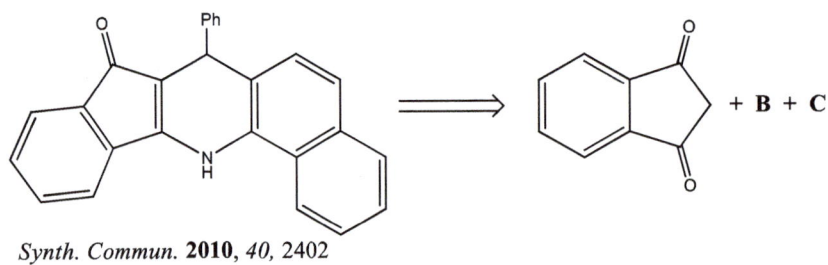

$$\Longrightarrow \quad + \; \mathbf{B} + \mathbf{C}$$

Synth. Commun. **2010**, *40,* 2402

11)

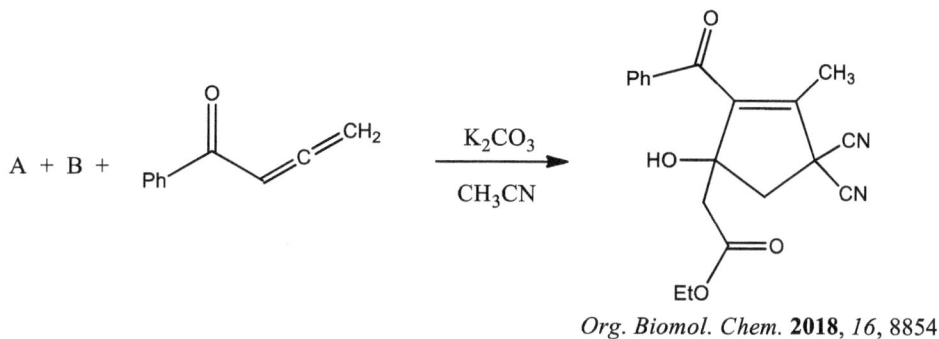

Org. Biomol. Chem. **2018**, *16*, 8854

12)

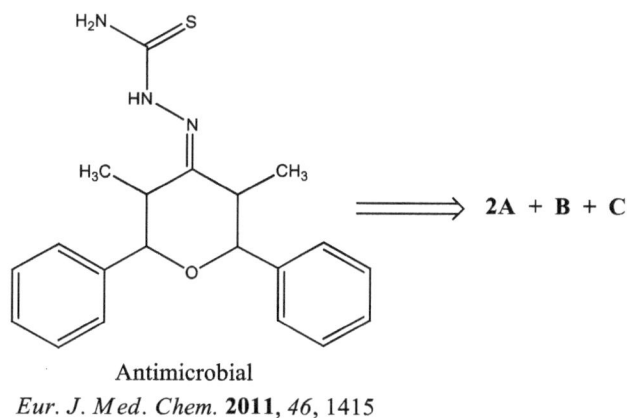

Antimicrobial
Eur. J. Med. Chem. **2011**, *46*, 1415

13)

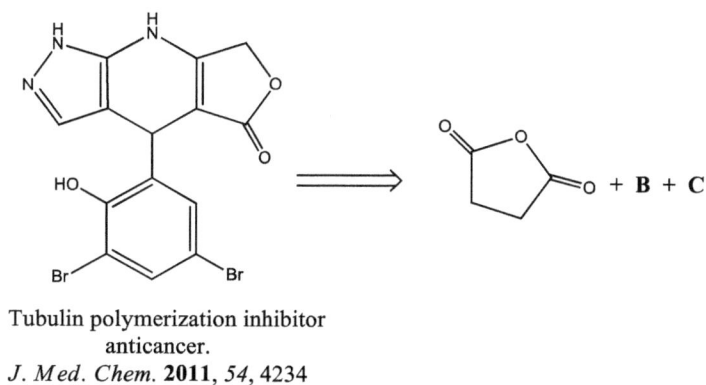

Tubulin polymerization inhibitor
anticancer.
J. Med. Chem. **2011**, *54*, 4234

14)

A + B + C + D $\xrightarrow[\text{rt-120°C}]{\text{MeOH}}$

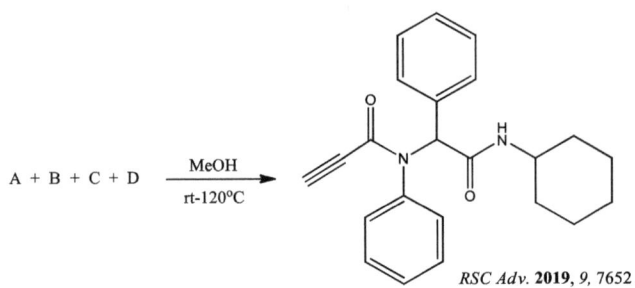

RSC Adv. **2019**, *9*, 7652

15)

A + B + C + D $\xrightarrow[\text{rt}]{\text{MeOH}}$

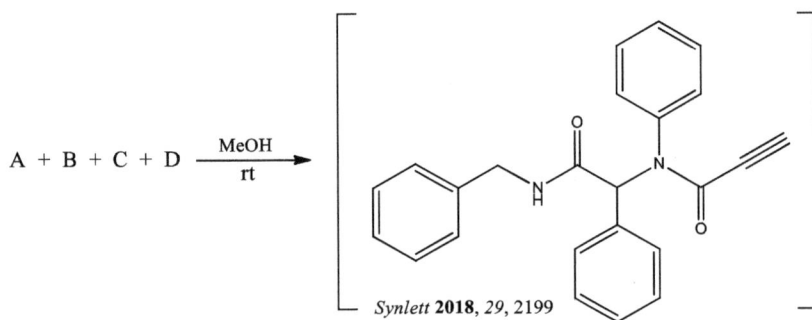

Synlett **2018**, *29*, 2199

16)

A + B + C + D $\xrightarrow[\text{rt, 2d, 83\%}]{\text{MeOH}}$

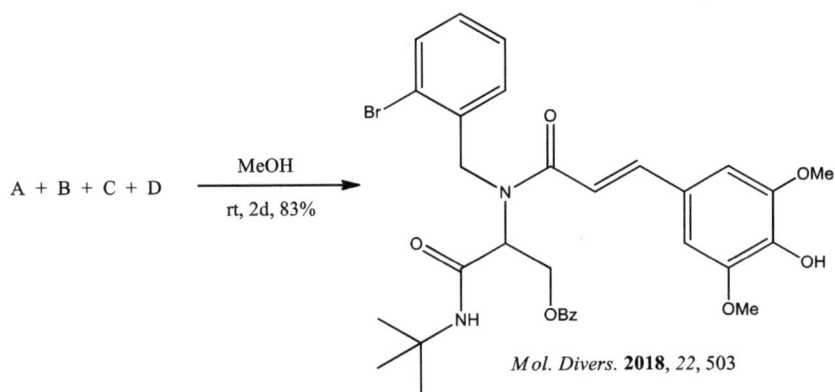

Mol. Divers. **2018**, *22*, 503

17)

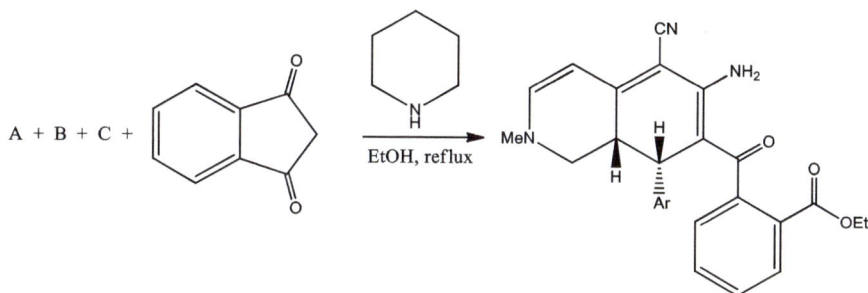

ChemistrySelect **2019**, *4*, 2663

18) In the exercise below, shows the synthesis of 1-tetrazolylimidazo[1,--a]pyridines based on Ugi multicomponent reaction. Propose the mechanism for the cyclic-acylation step for the formation of a series of tetrazole-linked imidazo[1,5-*a*] pyridines.

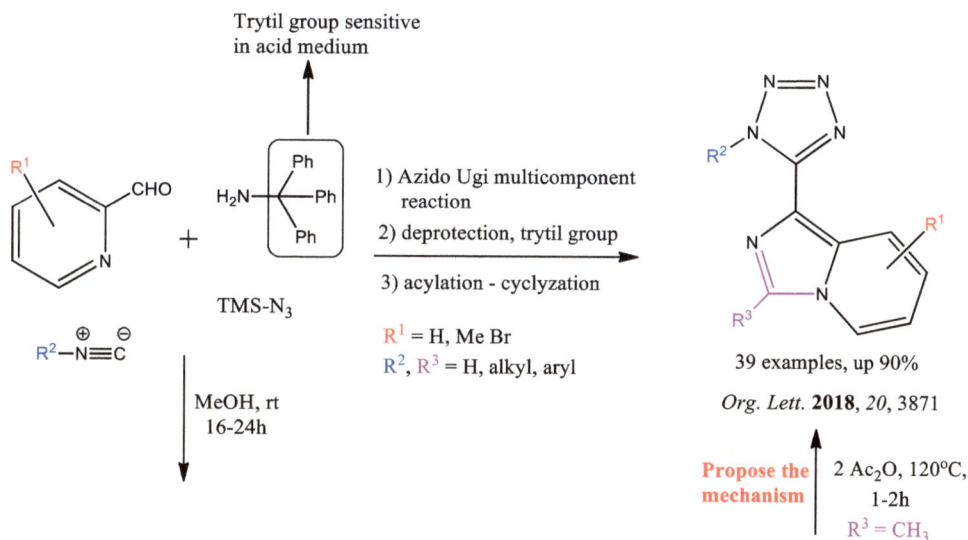

Trytil group sensitive in acid medium

1) Azido Ugi multicomponent reaction
2) deprotection, trytil group
3) acylation - cyclyzation

R^1 = H, Me Br
R^2, R^3 = H, alkyl, aryl

TMS-N_3

MeOH, rt
16-24h

39 examples, up 90%

Org. Lett. **2018**, *20*, 3871

Propose the mechanism | 2 Ac$_2$O, 120°C, 1-2h
R^3 = CH$_3$

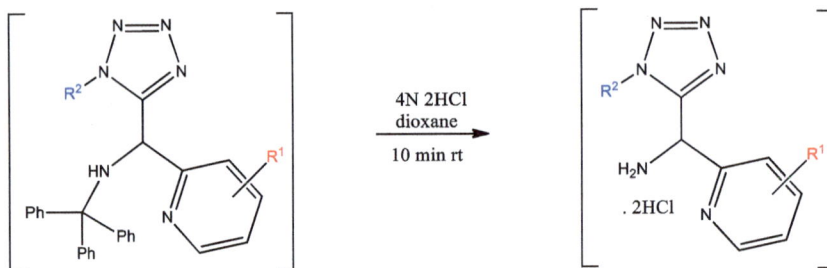

Answer

1)

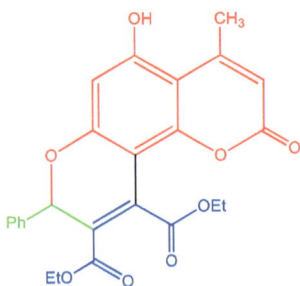

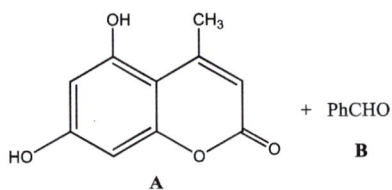

2) A = acetophenone; **B** = R^{1}CHO; **C** = amine (Mannich type reaction).

3)

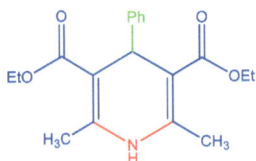

Hantzsch multicomponent reaction

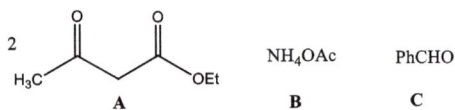

4)

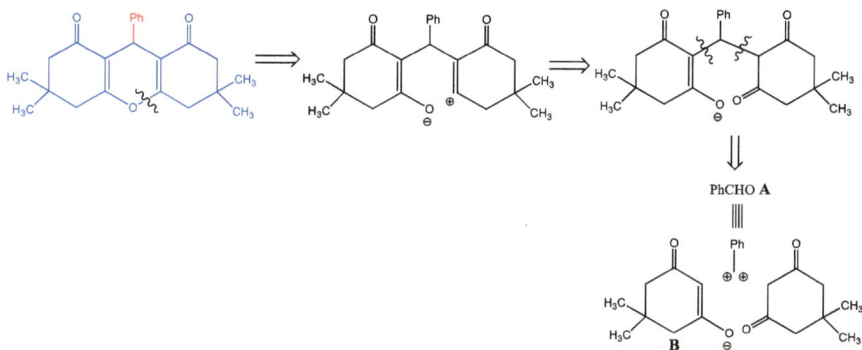

5)

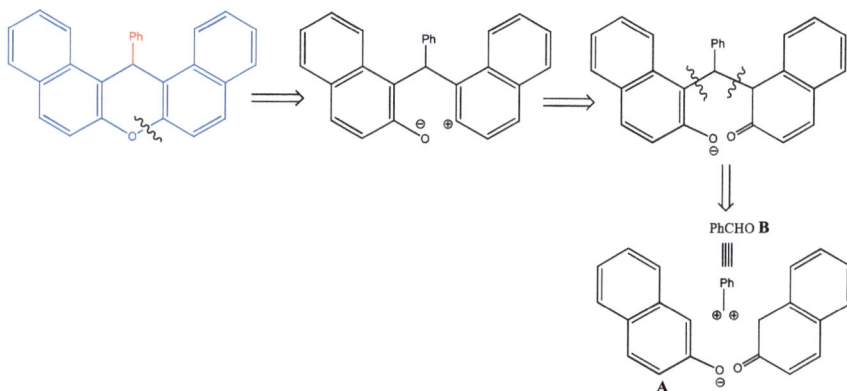

6)

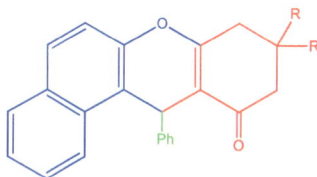

A = β-naphthol; **B** = Ph-CHO; **C** = 5,5-dimethyl-1,3-cyclohexanedione.

7)

Passerini reaction

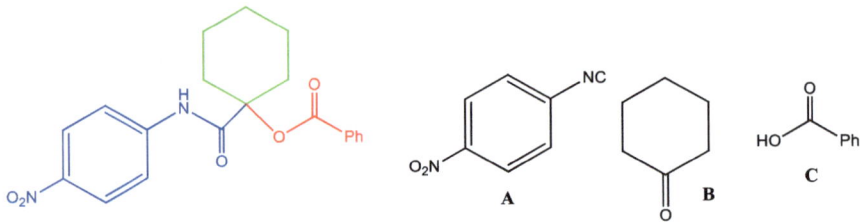

8)

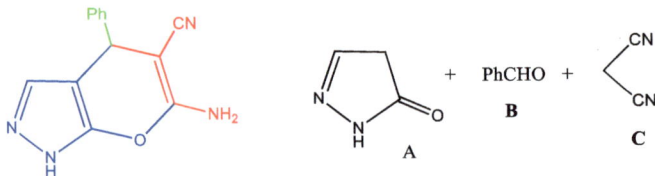

9)

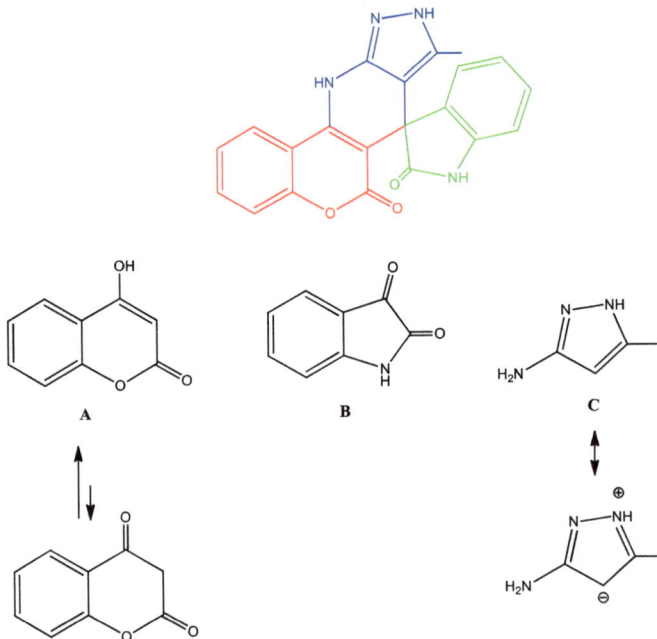

10)

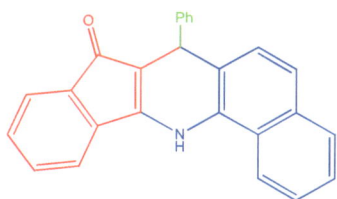   **B** ArCHO **C**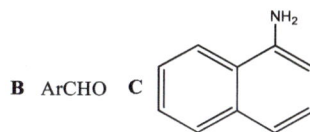

11)

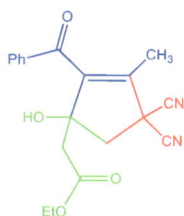   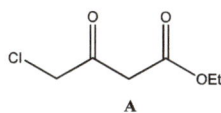

Plausible simplified mechanism

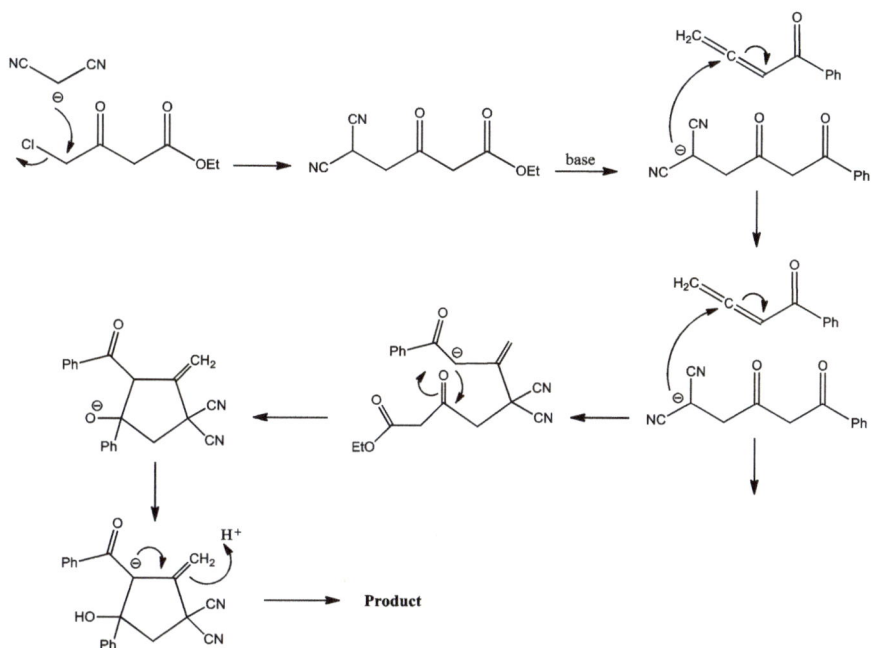

12)

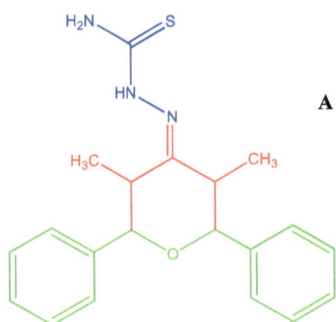   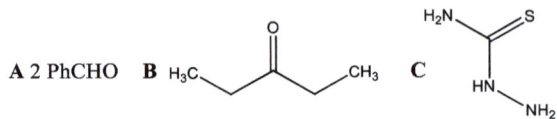

13)

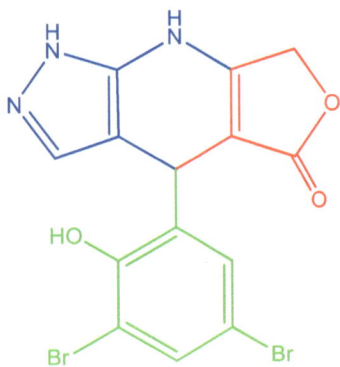   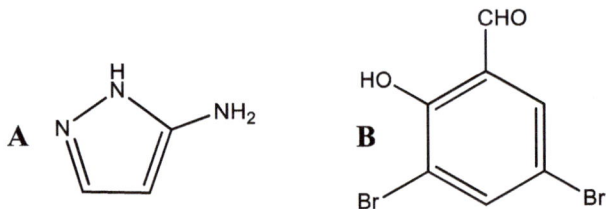

14)

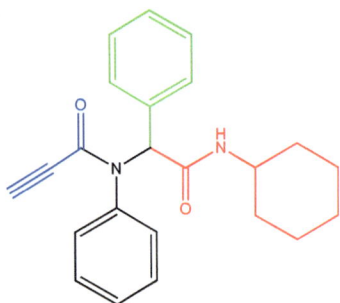   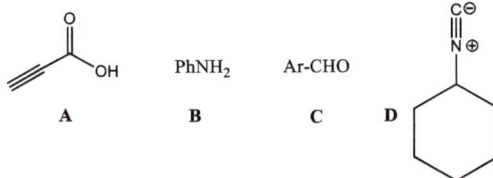

15)

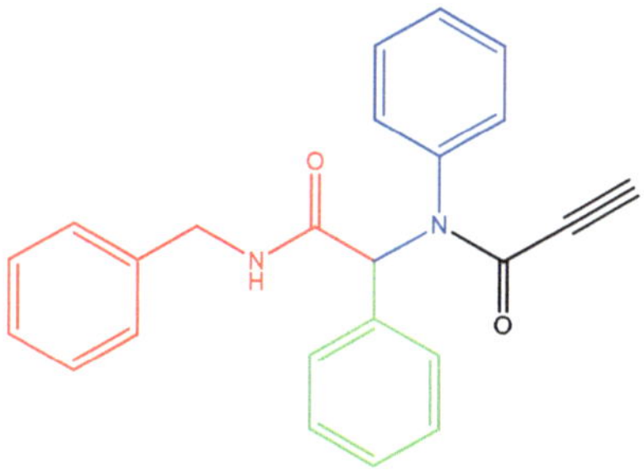

Ugi multicomponent reaction

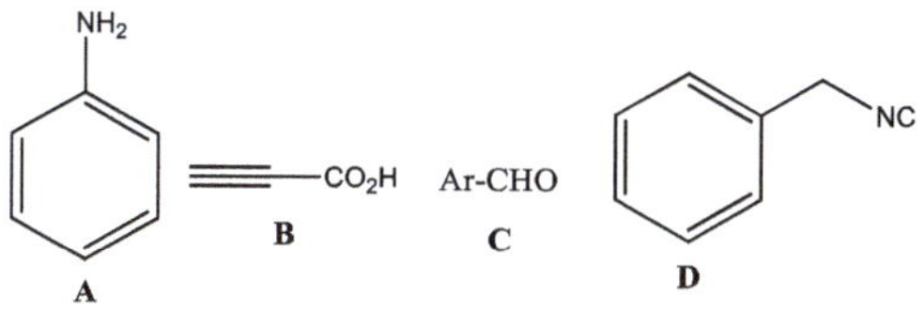

16)

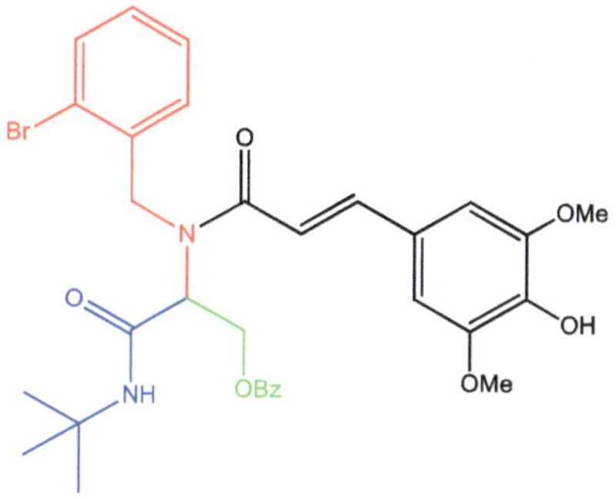

Ugi multicomponent reaction

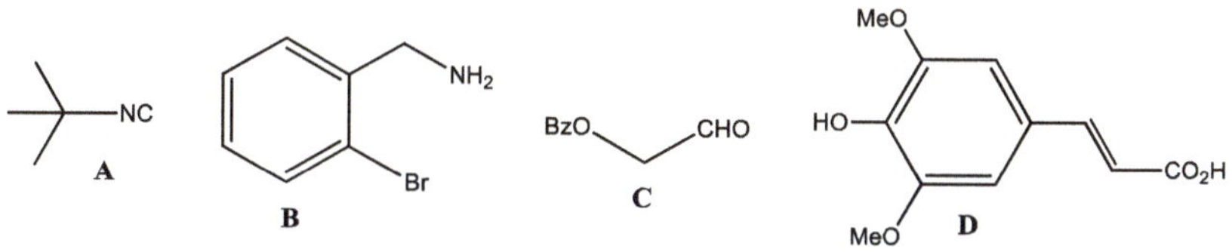

17)

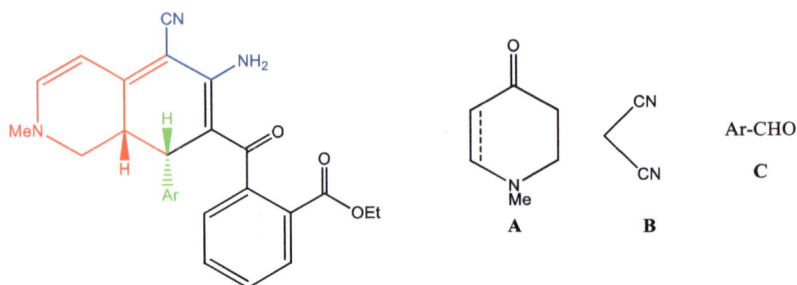

Plausible simplified mechanism

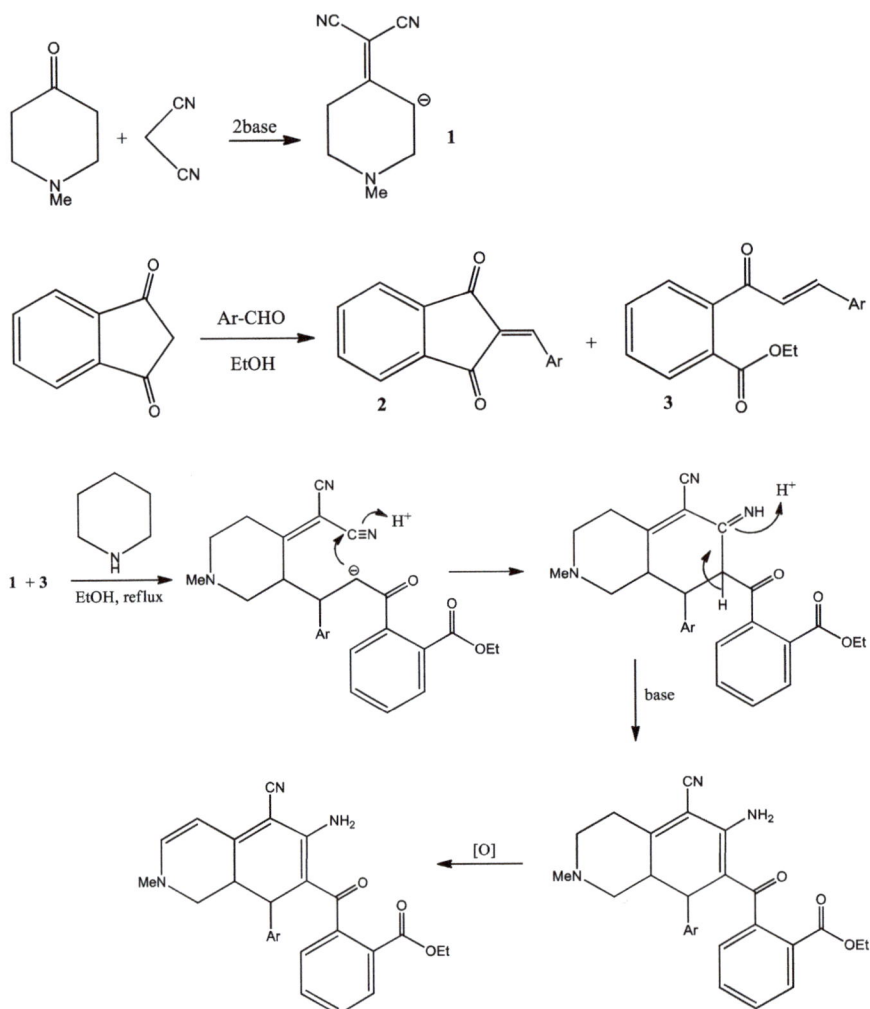

18)

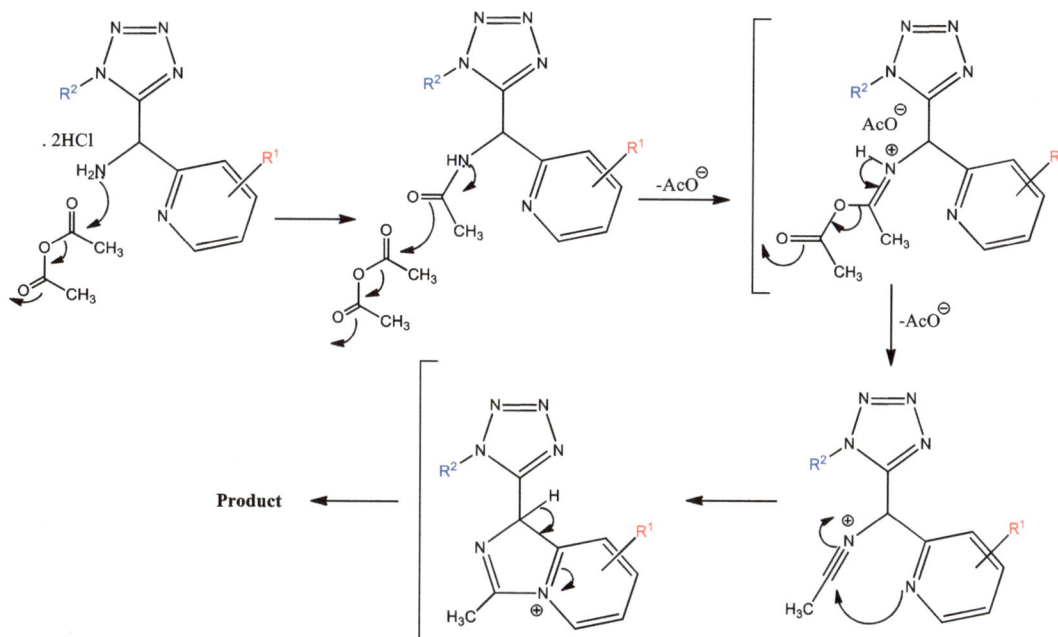

CONCLUSION

In organic synthesis, the reaction known as multicomponent reaction (MCR) is very useful in medicinal chemistry, producing a diversity of compounds for biological evaluation. This reaction leads to a mixture of three or more compounds or reagents, leading only to one or a major molecule being able to form complex molecules with a high degree of control of the stereogenic centers in only one step. In drug discovery, this is a type of reaction widely used nowadays, and the reader must know the topic.

REFERENCES

[1]   Ruijter, K.E.; Orru, R.V.A. Isocyanide-based multicomponent reactions towards cyclic constrained peptidomimetics. *Beilstein J. Org. Chem.,* **2014**, *10*, 544. https://www.beilstein-journals.org/bjoc/content/pdf/1860-5397-10-50.pdf

[2]   Rocha, R.O.; Rodrigues, M.O.; Neto, B.A.D. Review on the Ugi Multicomponent Reaction Mechanism and the Use of Fluorescent Derivatives as Functional Chromophores. *ACS Omega,* **2020**, *5*, 972. https://pubs.acs.org/doi/10.1021/acsomega.9b03684

[3]   Subramaniapilla, S.G.; Greene, T.W. Mannich reaction: A versatile and convenient approach to bioactive skeletons. *J. Chem. Sci.,* **2013**, *125*, 467. https://link.springer.com/article/ 10.1007/s12039-013-0405-y]

CHAPTER 9

Click Chemistry

Abstract: The most famous reaction, based on the click chemistry concept, is to obtain 1,2,3-triazoles, employing the 1,3-dipolar cycloaddition reaction between azides and alkynes. This reaction raised intense interest in academics and industrialists and was discovered independently by Sharpless and the Danish Morten Meldal in the year 2002. They demonstrated that the addition of copper as a catalyst has several advantages, such as an increase in the speed of the reaction, regioselectivity, high yields, and reactions are easy to elaborate. Due to its importance in drug discovery, the exercises below are based on this concept.

Keywords: 1,2,3-triazoles, Alkynes, Azides, Bioactive compounds, Catalyst, Chemical transformations, Click chemistry, Combinatorial chemistry, Copper, Drugs, Exercises, Heteroaromatic compounds, Medicinal chemistry, Molecules, Reaction conditions, Reagents, Retrosynthetic analysis, Substances, Synthesis, Synthons, 1,3-dipolar cycloaddition.

INTRODUCTION

The "click chemistry" is a concept proposed the same year that researcher Karl Barry Sharpless won the Nobel Prize in chemistry, *i.e.*, in 2001. This concept resembles the way that mother nature synthesizes its substances, bringing together, in a fast and effective way, small molecules. It is also related to the principles of green chemistry and combinatorial chemistry to have libraries of substances using efficient and straightforward procedures. These procedures should have the following characteristics: reduced numbers of steps, simplicity, the non-use of solvents or water, the non-generation of non-toxic waste or residues, and the use of inexpensive and abundant starting materials. Sharpless, which described the principles and concepts of click chemistry for the first time in one of his articles, enumerated several reactions that follow these principles and concepts. These reactions are Diels-Alder, epoxides, and aziridines opening, dihydroxylation, Michael addition, the formation of heteroaromatic nuclei, hydrazones, amides, ethers, oximes, ureas, and thioureas (Fig. **1**) [1].

However, the most famous reaction, based on the click chemistry concept, is to obtain 1,2,3-triazoles employing the 1,3-dipolar cycloaddition reaction between

Marcus Vinícius Nora de Souza

azides and alkynes (Scheme **1**). This reaction raised intense interest in both academics and industrialists and was discovered independently by Sharpless and the Danish Morten Meldal in 2002. They demonstrated that the addition of copper as a catalyst has several advantages, such as an increase in the speed of the reaction, regioselectivity, high yields, and reactions are easy to elaborate. The catalytic cycle of this reaction is represented in Scheme **2**. If the reader is interested in the topic, below are some suggestions for reading [2 - 7].

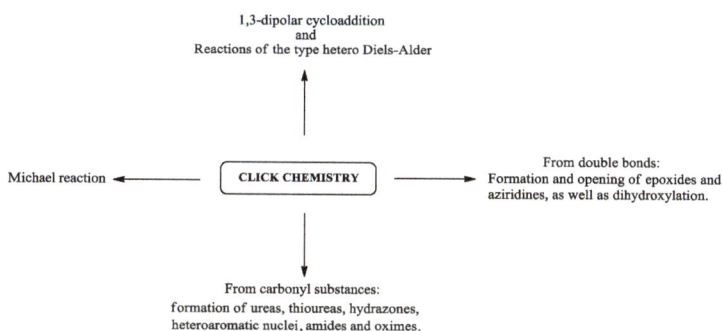

Fig. (1). Reactions related to the concept of click chemistry.

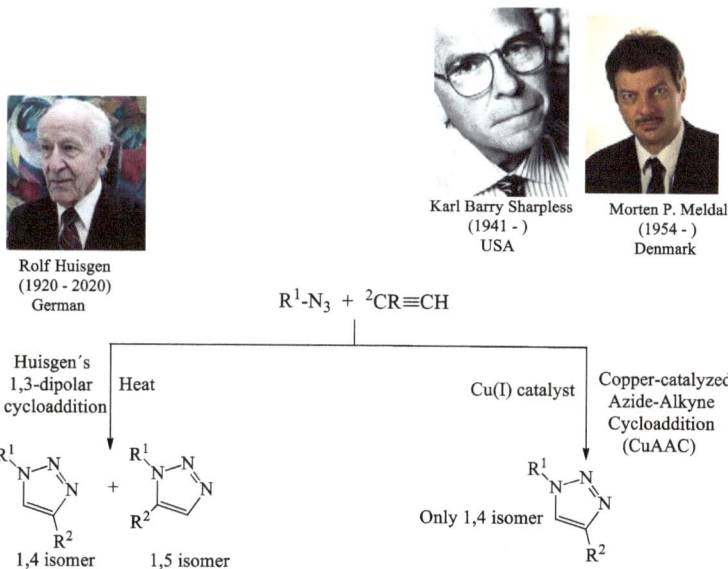

Scheme 1. 1,2,3-triazole formation by Huisgen's reaction and by regioselective copper-catalyzed azide-alkyne cycloadditions (CuAAC).
Source of the picture: Wikipedia
https://en.wikipedia.org/wiki/Rolf_Huisgen
Source of the picture: The Nobel Prize website.
https://www.nobelprize.org/prizes/chemistry/2001/sharpless/biographical/
Source of the picture: University of Copenhagen website.
https://synbio.ku.dk/about/management/ulrik_gether_copy/

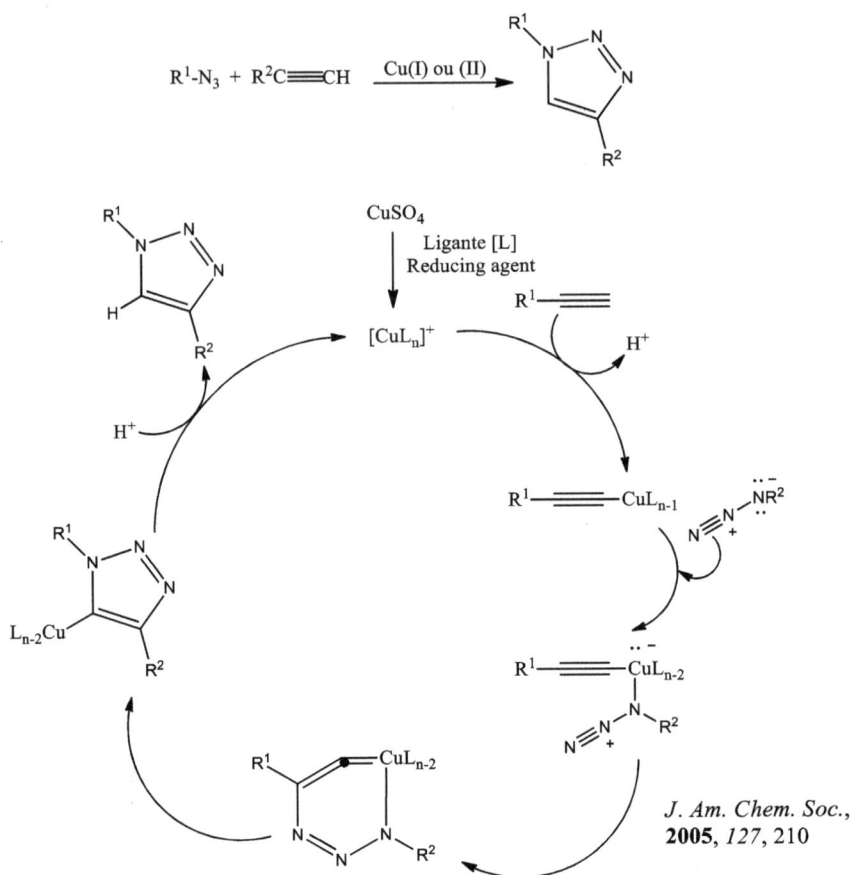

Scheme 2. Catalytic cycle of 1,3-dipolar cycloaddition reaction between azides and alkynes to form 1,2,3 triazoles.

Example

a) Based on the two substances below, benzyl 2,6-difluorobromide and propionic acid, how would the reader synthesize the antiepileptic rufinamide using the "click chemistry" reaction?

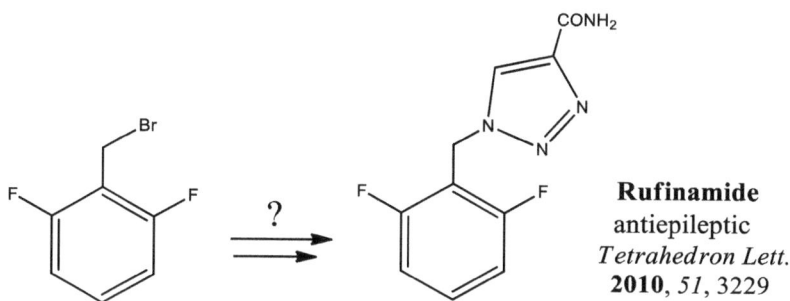

Rufinamide
antiepileptic
Tetrahedron Lett.
2010, *51*, 3229

Answer

a) Based on the target molecule (retrosynthetic analysis), it is easy to identify the transformations necessary to obtain this drug.

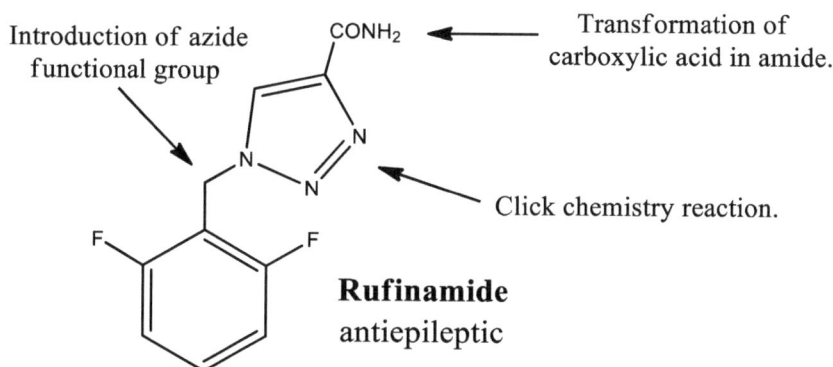

Introduction of azide functional group

Transformation of carboxylic acid in amide.

Click chemistry reaction.

Rufinamide
antiepileptic

(a) NaN_3; **(b)** Cu(I), base, MeCN and propionic acid ($HCCCO_2H$); **(c)** $SOCl_2$; **(d)** NH_3.

EXERCISES

Due to the importance of the "click chemistry" concept in the development of new drugs, the reader is invited to solve the following exercises using this type of reaction as a crucial step for the synthesis of 1,2,3-triazoles.

In exercises **1** through **6** below, complete the gaps for the synthesis of their respective compounds.

1)

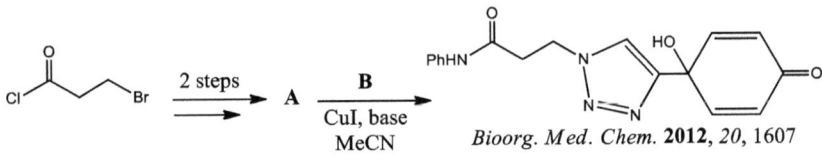

Bioorg. Med. Chem. **2012**, *20*, 1607

2)

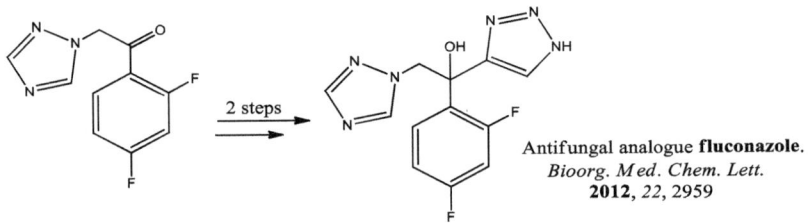

Antifungal analogue **fluconazole**.
Bioorg. Med. Chem. Lett.
2012, *22*, 2959

3)

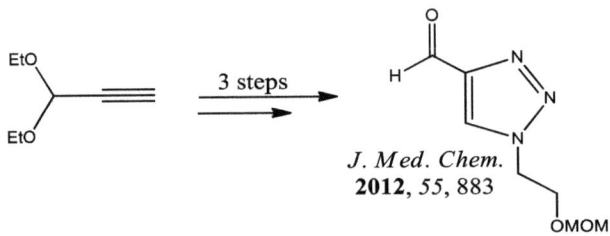

J. Med. Chem.
2012, *55*, 883

4)

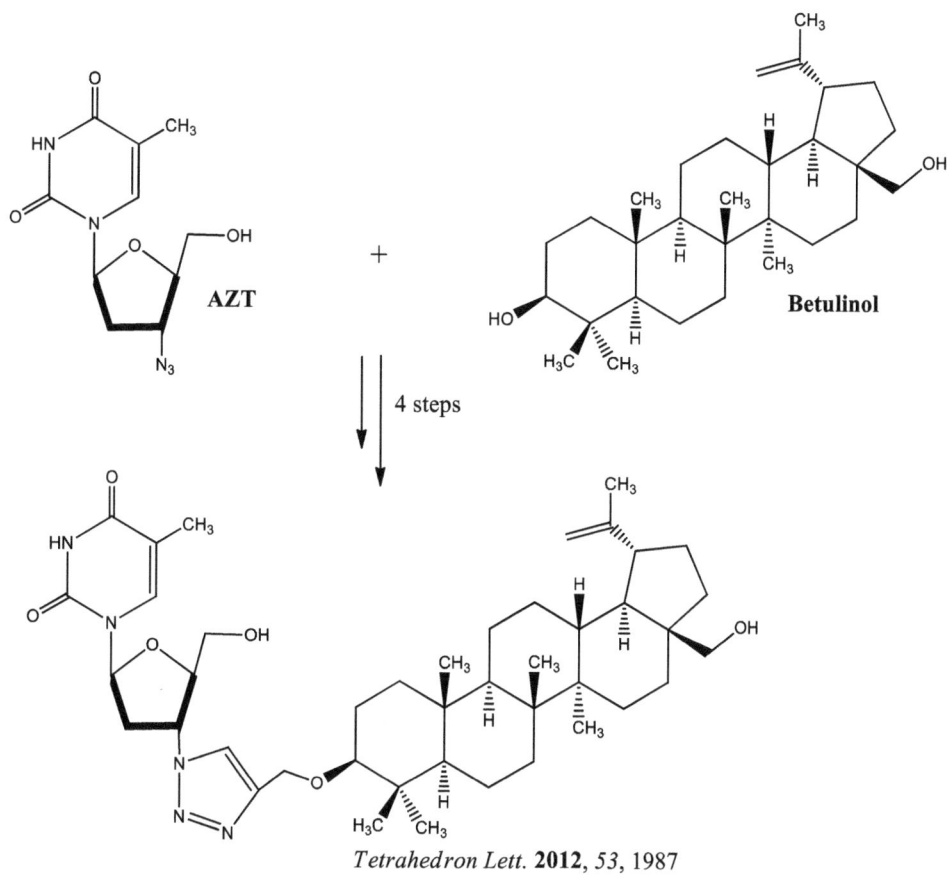

Tetrahedron Lett. **2012**, *53*, 1987

5)

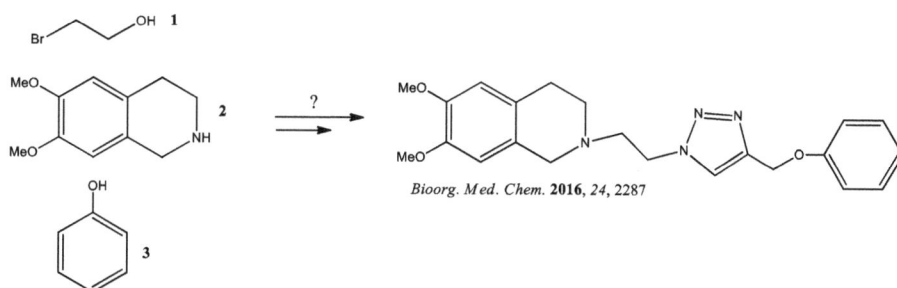

Bioorg. Med. Chem. **2016**, *24*, 2287

6)

$$A + B \xrightarrow[\substack{\text{DMF-H}_2\text{O (8:2)} \\ 25\text{-}40^\circ\text{C, 6-10h}}]{\substack{\text{NaN}_3,\ \text{CuSO}_4.5\text{H}_2\text{O} \\ \text{sodium ascorbate}}}$$

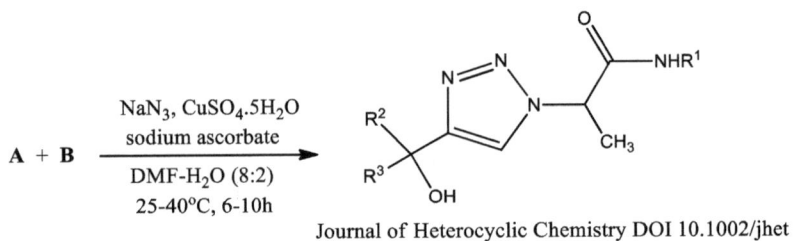

Journal of Heterocyclic Chemistry DOI 10.1002/jhet

In exercises **7** through **9** below use click chemistry as a critical reaction in the synthesis of the target molecules.

7)

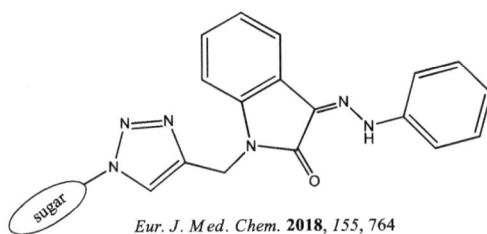

Eur. J. Med. Chem. **2018**, *155, 764*

8)

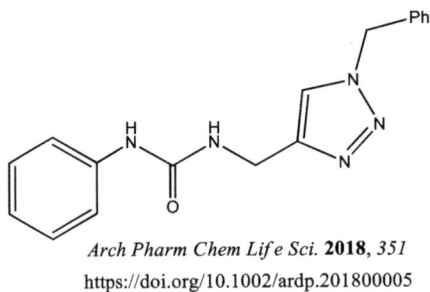

Arch Pharm Chem Life Sci. **2018**, *351*
https://doi.org/10.1002/ardp.201800005

9)

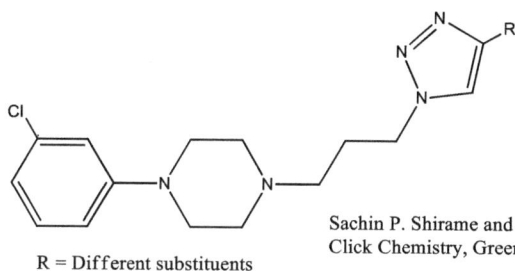

R = Different substituents

Sachin P. Shirame and Raghunath B. Bhosale. Green Approach in Click Chemistry, Green Chemistry, 2018, pg. 175, IntechOpen.

10) In this click chemistry reaction, copper was not used as the catalyst producing two compounds. Which ones are they?

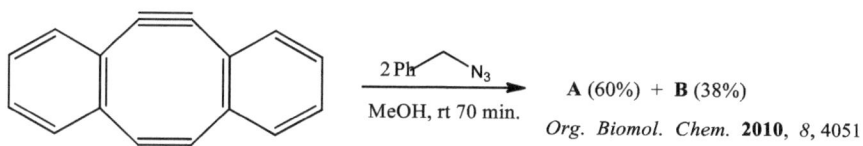

2 Ph⌒N₃

MeOH, rt 70 min.

A (60%) + **B** (38%)

Org. Biomol. Chem. **2010**, *8*, 4051

Answers

1)

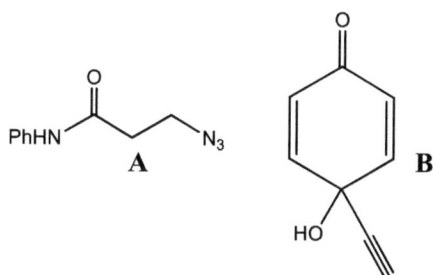

2)

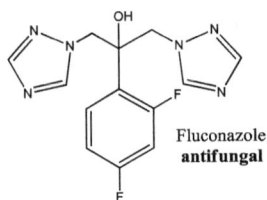

(**a**) Zn, propargyl bromide;
(**b**) CuI, NaN₃, sodium ascorbate, CuSO₄, DMSO.

Fluconazole
antifungal

3) (a) N₃CH₂CH₂OH; sodium ascorbate, CuSO₄.5H₂O, *t*-BuOH, water, rt, 4h; **(b)** TFA (50%), CHCl₃, 0°C, 2h; **(c)** MOMCl, DIPEA, THF, 65°C, 16h.

4) (a) TBSCl (2.0 eq.) regioselective protection of the primary hydroxyls of the two starting materials (AZT and betulinol); **(b)** base followed by propargyl bromide; **(c)** click chemistry; **(d)** 2TBAF (deprotection of silyl groups).

5) 1(i) NaN₃, water, 80°C, 24 h; **(ii)** TEA/DCM, TsCl, rt, 24 h;

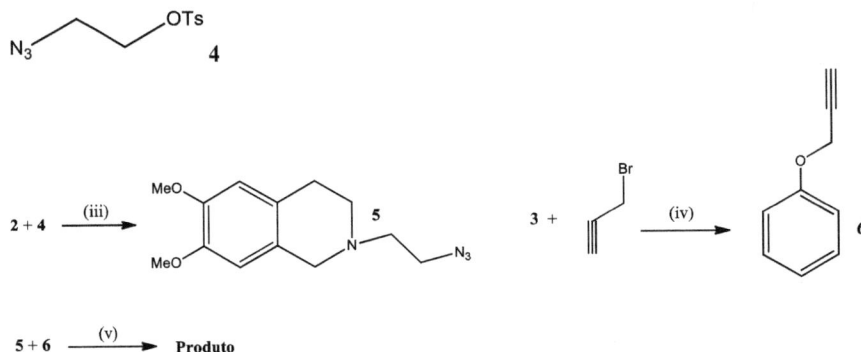

(iii) TEA/acetonitrile, 60 °C, 24 h; **(iv)** K₂CO₃, acetone, reflux, 4 h; **(v)** ascorbate sodium, CuSO₄, 75% CH₃OH, 24–48 h.

6)

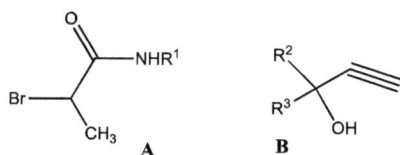

7)

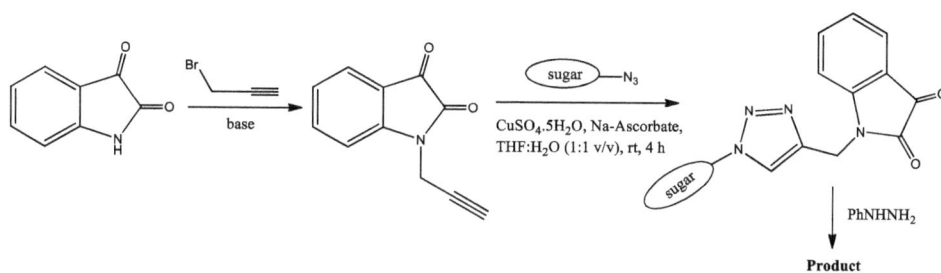

8)

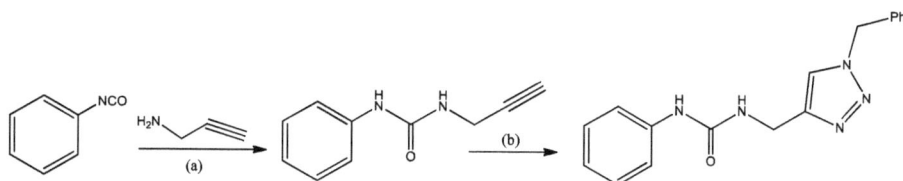

(a) CH_2Cl_2, 20°C; (b) (i) $ArCH_2Br(Cl)$, NaN_3, H_2O/t-BuOH (1:1), Et_3N; (ii) 3, $CuSO_4$, sodium ascorbate, rt

9)

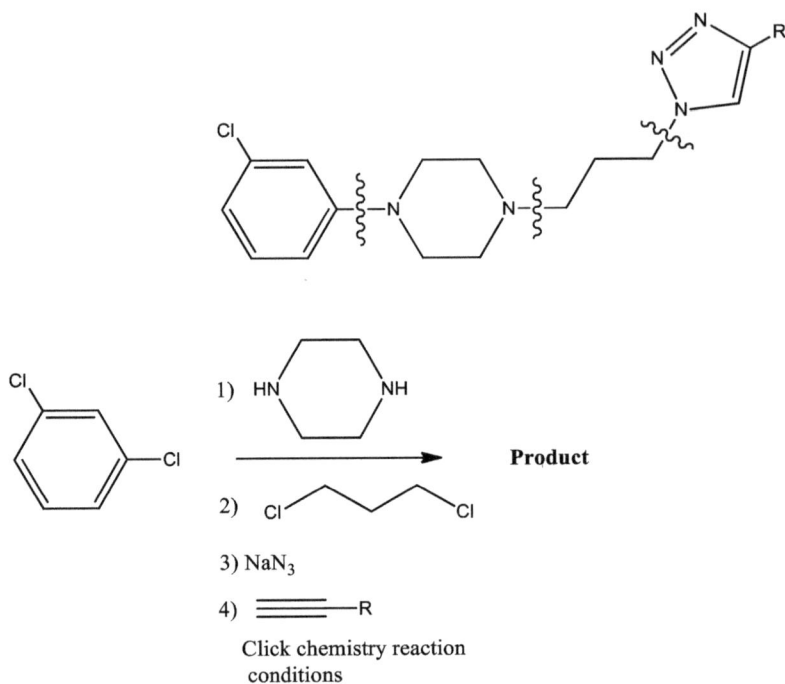

1) HN—NH (piperazine)

2) Cl⌒⌒Cl

3) NaN₃

4) ☰—R

Click chemistry reaction conditions

→ **Product**

10)

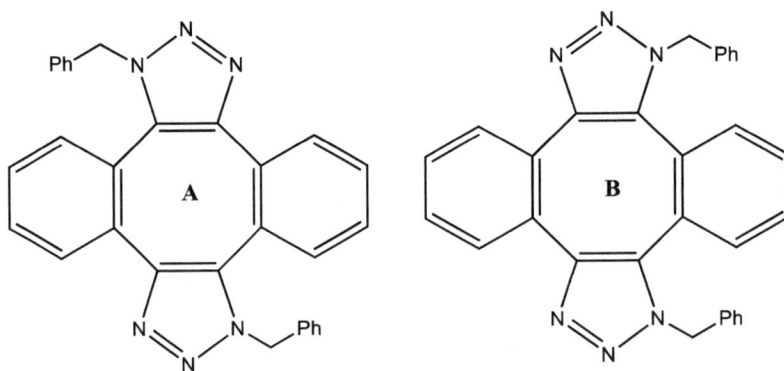

A   B

CONCLUSION

A click chemistry concept is a useful tool in organic synthesis and medicinal chemistry. This concept resembles how mother nature synthesizes its substances, bringing together, in a fast and effective way, small molecules. It is also related to the principles of green chemistry and combinatorial chemistry to have libraries of substances using efficient and straightforward procedures. The most famous reaction, based on the click chemistry concept, is to obtain 1,2,3-triazoles employing the 1,3-dipolar cycloaddition reaction between azides and alkynes. This reaction is an important tool in the drug discovery.

REFERENCES

[1]   Kolb, H.C.; Finn, M.G.; Sharpless, K.B. Click Chemistry: Diverse Chemical Function from a Few Good Reactions. *Angew. Chem. Int. Ed.,* **2001**, *40*(11), 2004-2021.
[http://dx.doi.org/10.1002/1521-3773(20010601)40:11<2004::AID-ANIE2004>3.0.CO;2-5]

[2]   Rostovtsev, V.V.; Green, L.G.; Fokin, V.V.; Sharpless, K.B. A stepwise huisgen cycloaddition process: copper(I)-catalyzed regioselective "ligation" of azides and terminal alkynes. *Angew. Chem. Int. Ed.,* **2002**, *41*(14), 2596-2599.
[http://dx.doi.org/10.1002/1521-3773(20020715)41:14<2596::AID-ANIE2596>3.0.CO;2-4]

[3]   Himo, F.; Lovell, R.; Hilgraf, R.; Rostovtsev, V.V.; Noodleman, L.; Sharpless, K.B.; Fokin, V. Copper(I)-Catalyzed Synthesis of Azoles.. DFT study predicts unprecedented reactivity and intermediates. *J. Am. Chem. Soc.,* **2005**, *127*(21), 210-216.
[http://dx.doi.org/10.1021/ja0471525] [PMID: 15631470]

[4]   Freitas, L.B.O.; Ruela, F.A.; Pereira, G.R.; Brondi, R.A.; Freitas, R.P.; Santos, L.S. The "click" reaction in the synthesis of 1,2,3-triazoles: chemical aspects and applications. *Quim. Nova,* **2011**, *34*, 1791-1804.
[http://dx.doi.org/10.1590/S0100-40422011001000012]

[5]   Christopher, D.H.; Xin-Ming, L.; Dong, W. Click chemistry, a powerful tool for pharmaceutical sciences. *Pharm. Res.,* **2008**, *25*(10), 2216-2230.
[http://dx.doi.org/10.1007/s11095-008-9616-1] [PMID: 18509602]

[6]   Kaushik, C.P.; Sangwan, J.; Luxmi, R.; Kumar, K.; Pahwa, A. Synthetic Routes for 1,4-disubstituted 1,2,3-triazoles: A Review. *Curr. Org. Chem.,* **2019**, *23*(8), 860-900.
[http://dx.doi.org/10.2174/1385272823666190514074146]

[7]   De Souza, M.V.; Franca, C.C.; Facchinetti, V.; Brandao, C.G.; Pacheco, P.M. Advances in triazole synthesis from copper-catalyzed azide-alkyne cycloadditions (CuAAC) based on eco-friendly procedures. *Curr. Org. Synth.,* **2019**, *16*(2), 244-257.
[http://dx.doi.org/10.2174/1570179416666190104141454] [PMID: 31975674]

CHAPTER 10

Fluorine in Drug Discovery

Abstract: Fluorine chemistry is an important area that made a significant impact on drug discovery development, which can be seen by a large number of drugs in the market containing this chemical element. The success of drugs containing fluorine atoms in its structures is due to its chemical properties compared to the other elements, providing several advantages in drug discovery. Some of these advantages are better metabolic stability, avoiding undesired metabolizations, and improving the bioavailability of the drugs, among others. Due to its importance in drug discovery, the exercises below are focused on fluorine chemistry.

Keywords: Bioactive compounds, Catalyst, Chemical transformations, Combinatorial chemistry, Drugs, Exercises, Fluorine, Heteroaromatic compounds, Medicinal chemistry, Molecules, Reaction conditions, Reagents, Retrosynthetic analysis, Substances, Synthesis, Synthons.

INTRODUCTION

Fluorine chemistry is an important area that made a significant impact on drug discovery development, which can be seen by a large number of drugs in the market containing this chemical element (Fig. **1**) [1, 2].

The success of drugs that contain fluorine atoms in its structures and also in other areas, such as materials, agrochemicals, and radiotracers for positron emission tomography (PET), is due to its chemical proprieties compared to the other elements (Table **1**) [3]. The C-F bond (105 Kcal/mol) is stronger than the C-H bond (98 Kcal/mol), providing several advantages on drug discovery. Some of these advantages are better metabolic stability, avoiding undesired metabolizations, and improving the bioavailability of the drugs. The size of the fluorine atom (van der Waals radius 1.47 Å) is close to oxygen (1.52 Å) and hydrogen (1.2 Å) being able to mimic the last one in medicinal chemistry [3]. A fluorine atom is the most electronegative element (4 Pauling units) of the periodic table. Its presence in a molecule provides the ability to modify a series of parameters. Some of these parameters are conformational changes, variations of pKa, formation of hydrogen bonds, and the increase and decrease in lipophilicity.

Marcus Vinícius Nora de Souza

These characteristics can change both the pharmacodynamics and pharmacokinetic parameters of a bioactive compound (Scheme **1**) [4].

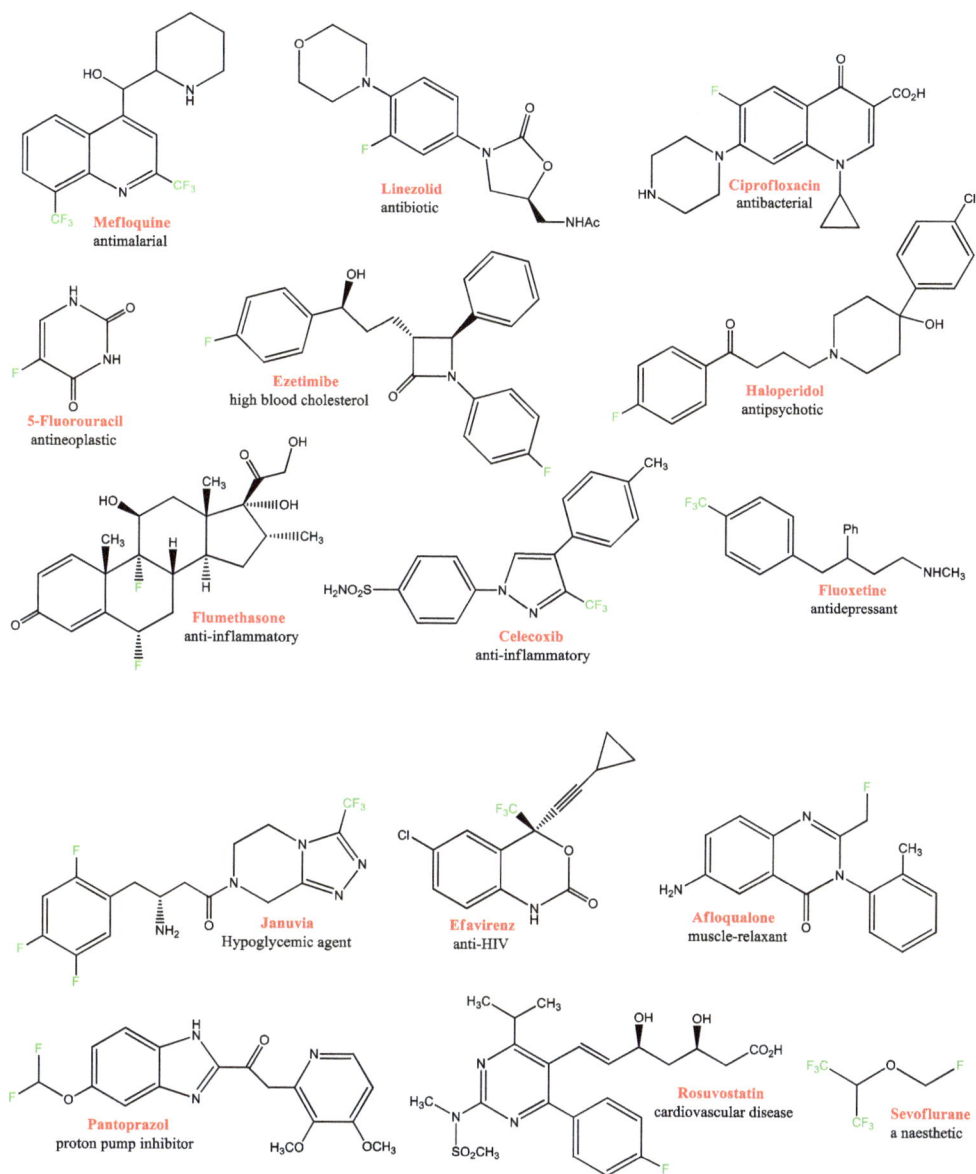

Fig. (1). Drugs that contain fluorine atoms in its structures.

Table 1. Comparison of the fluorine atom with other atoms.

	H	C	N	O	F	Cl
Bond strength to C (Kcal/mol)	98	83	70	84	105	77
Bond length to C (Å)	1.09	1.53	1.47	1.43	1.39	1.76
Van der Waals radius (Å)	1.2	1.7	1.55	1.52	1.47	1.75
Electronegativity (Pauling units)	2.1	2.5	3	3.5	4	3

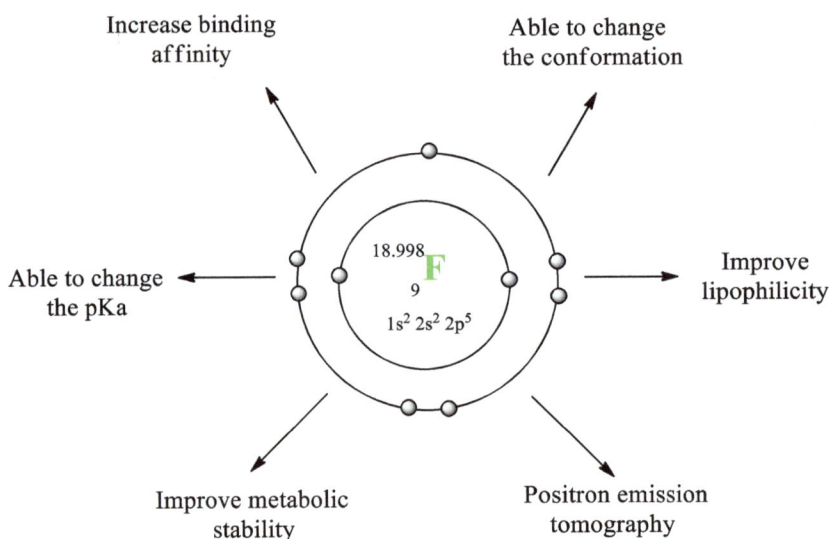

Scheme 1. The versatility of the fluorine atom in medicinal chemistry.

FLUORINATION REAGENTS

Due to the importance of fluorine atoms in organic molecules, different reagents have been developed. The classification of these reagents is based on nucleophilic and electrophilic fluorination, which are for the introduction of fluorine atom (-F), difluoromethyl (-CF_2H), and trifluoromethyl (-CF_3) group. Below are the most commonly used reagents for fluorination.

Nucleophilic Fluorination for the Introduction of Fluorine Atom NaF, KF, HF, and Bu₄NF

For fluorination, using fluoride ion (F-) as a nucleophile, the commonly used reagents are sodium fluoride (NaF), potassium fluoride (KF), hydrogen fluoride

(HF), and quaternary ammonium salts such as Bu$_4$NF (TBAF = tetra-*n*-butylammonium fluoride). Hydrogen fluoride is also known as Olah´s reagent, which is a mixture of 70% of hydrogen fluoride and 30% of pyridine used to transform alcohol in alkyl fluorides (Scheme **2**) [5].

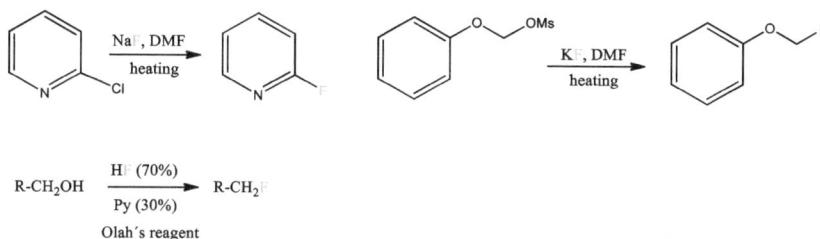

Scheme 2. Transformation of alcohols to alkyl fluorides.

Diethylaminosulfur Trifluoride (DAST)

Diethylaminosulfur trifluoride, known as **DAST** (Et$_2$NSF$_3$), is a useful reagent in fluorination, which is very often used to transform alcohols in their respective alkyl fluorides and also to produce geminal difluorides from aldehydes and unhindered ketones (Scheme **3**) [6].

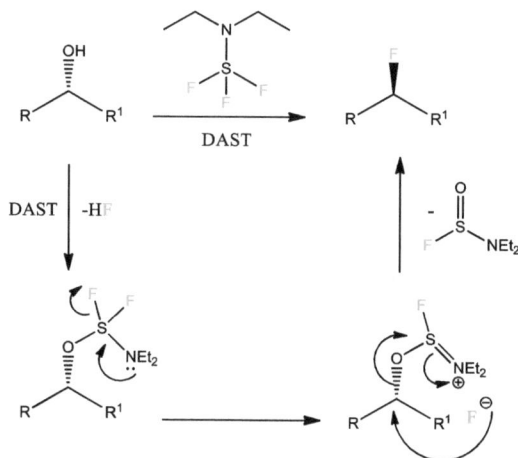

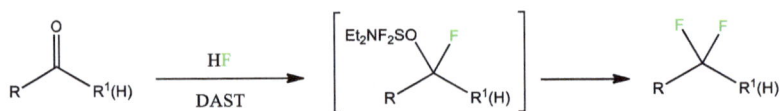

Scheme 3. Proposal mechanism of DAST.

Nucleophilic Fluorination for the Introduction of the Trifluoromethyl Group

Trifluoromethyltrimethylsilane (TMSCF$_3$)

The trifluoromethyl is an important group presented in several modern drugs and due to its importance, several methodologies have been developed. A very useful and versatile reagent for nucleophilic trifluoromethylation is trifluoromethyl-trimethylsilane (TMSCF$_3$), known as Ruppert-Prakash reagent, which is based on a pronucleophile of trifluoromethyl anion, as is described in Scheme 4 [7].

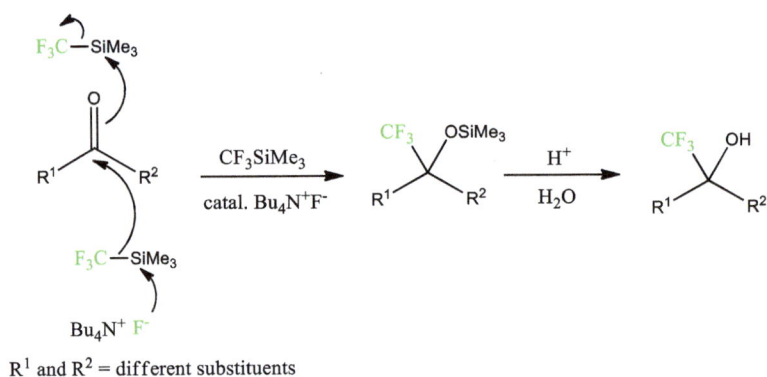

R^1 and R^2 = different substituents

Scheme (4). Mechanism of trifluoromethyltrimethylsilane (TMSCF$_3$) as nucleophilic trifluoromethylation.

Eletrophilic Fluorination for the Introduction of Fluorine Atoms

For the introduction of fluorine atoms based on electrophilic fluorination, there are three commonly used reagents: selectfluor, *N*-fluoro-benzenedisulfonimide, and *N*-fluorobenzenesulfonimide, which are focused on N-F reagents (Scheme 5). The mechanism of using these reagents is based on two proposals S$_N$2 and single-electron transfer (SET) (Scheme 6).

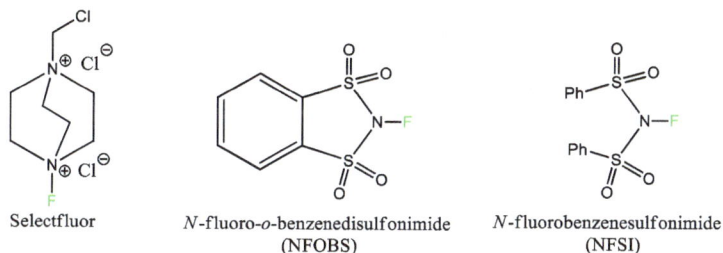

Scheme 5. Electrophilic fluorination reagents for the introduction of fluorine.

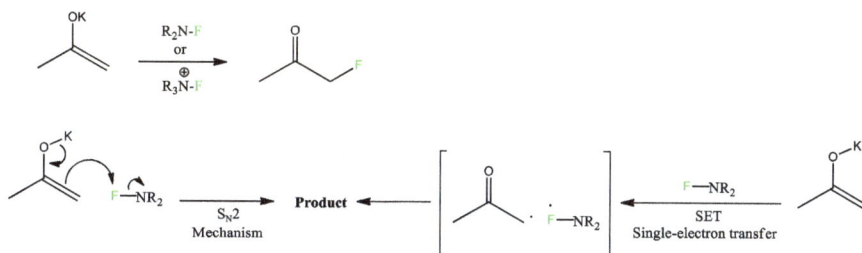

Scheme 6. Proposal mechanism of electrophilic fluorination using selectfluor, NFBOs, and NFSI.

Electrophilic Fluorination for the Introduction of the Trifluoromethyl Group

In the case of eletrophilic fluorination, Togni reagent (3,3-Dimethyl-1-(trifluoromethyl)-1,2-benziodoxole), 5-(trifluoromethyl) dibenzothiophenium-trifluoro methanesulfonate, and 5-(trifluoromethyl) dibenzothiophenium tetraborate are the most common (Scheme **7**) [8].

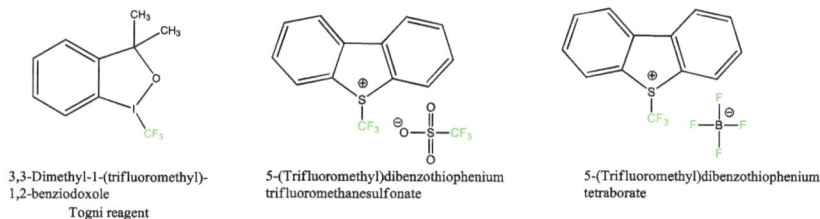

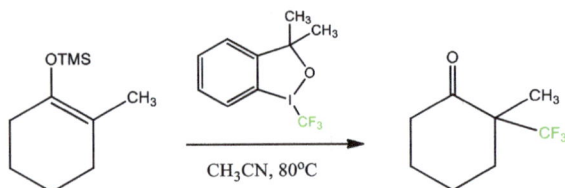

Scheme 7. Most common electrophilic fluorination reagents for the introduction of the trifluoromethyl group and its application.

Radical Fluorination

In addition to nucleophilic and electrophilic fluorination, there is also a free radical approach responsible for generating fluorine radical species. For example, Li and co-workers, by using aliphatic carboxylic acids, developed silver-catalyzed decarboxylative fluorination with silver nitrate ($AgNO_3$) as catalyst and selectfluor as a source of atomic fluorine radical (F·) (Scheme **8**) [8].

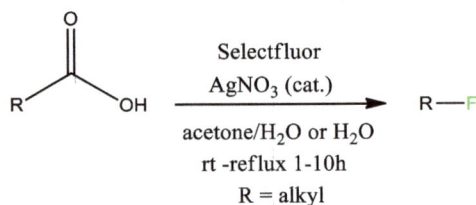

Scheme 8. Radical fluorination using Selectfluor.

Another decarboxylative fluorination was made by Sammis and co-workers with the thermolysis of *t*-butyl peresters to produce alkyl radicals, using NFSI or selectfluor as a source of atomic fluorine radical (F.). This methodology is useful for a wide variety of alkyl radicals, such as benzylic, heteroatom-stabilized radicals, primary, secondary, and tertiary radicals (Scheme **9**) [9, 10].

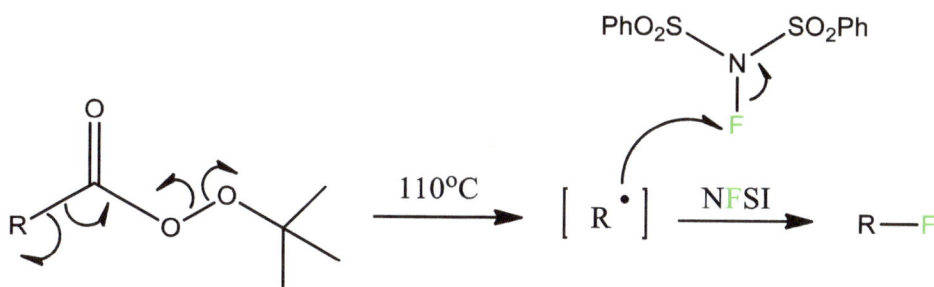

Scheme 9. Decarboxylative fluorination.

CLASSICAL EXAMPLES OF THE IMPORTANCE OF FLUORINE IN DRUG DEVELOPMENT

In the literature, there are classic examples of the use of fluorine in drug development, which are in several articles and books. Due to their importance and didacticism, these examples are below [4].

Fluoroquinolone

Fluoroquinolone is an essential class of antibacterials widely used nowadays. This class started with nalidixic acid (FDA - 1963) that, after a SAR (structure-activity relationship) study, found the fluoroquinolone norfloxacin (FDA – 1986). This drug opens a new era for this class, being 1000 times more potent than nalidixic acid (Scheme **10**) [11]. The reason for that is the combination of a fluorine atom at position **6** and the piperazinyl group at position **7** in the quinolone nucleus. The fluorine atom at this position is critical for the biological activity being responsible for improving the cell penetration and bacterial potency. This improvement is due to a better gyrase affinity, a vital enzyme that participates in the reparation, transcription, and replication of the bacterial DNA. The piperazinyl group was responsible for a better pharmacokinetic profile of this class. Another development was the introduction of the cyclopropyl group at position **1** in the fluoroquinolone nucleus being ciprofloxacin drug (FDA – 1994), the first of this class [4, 11].

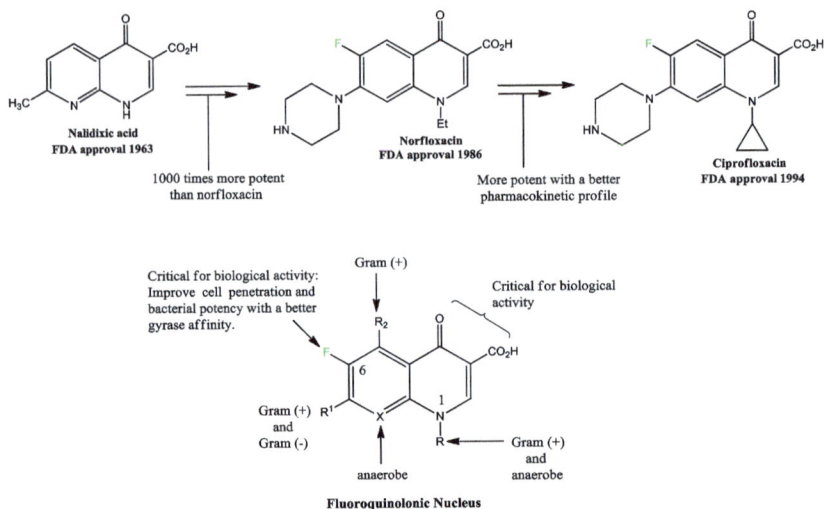

Scheme 10. SAR (structure-activity relationship) of fluoroquinolones.

Sitagliptin

Sitagliptin (FDA – 2006) is a drug developed by Merck & Co to treat diabetes mellitus type 2 by targeting the enzyme dipeptidyl peptidase 4 (DPP-4). The introduction of the fluorine atom and trifluoromethyl group was responsible for improving the potency, bioavailability, and absorption of this drug, demonstrating the importance of this atom in medicinal chemistry (Scheme **11**) [12].

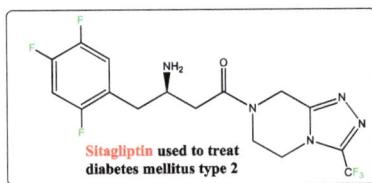

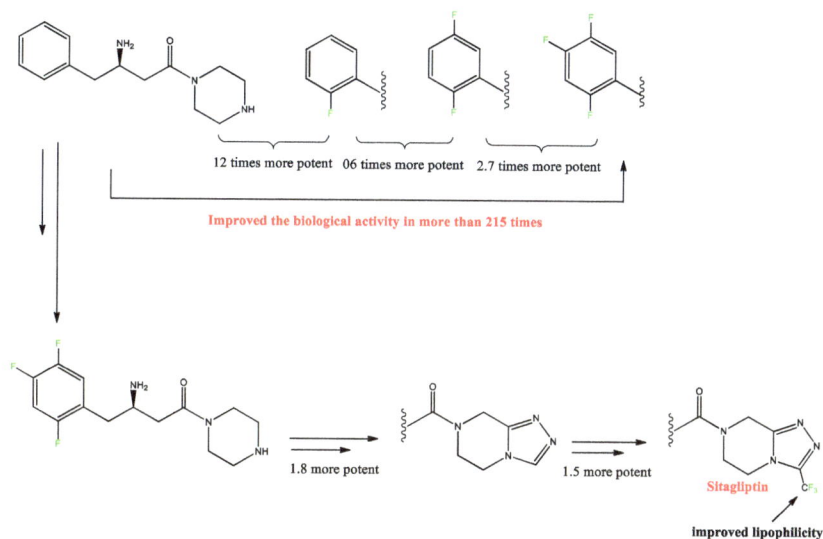

Scheme 11. Development of the drug Sitagliptin (FDA – 2006).

Odanacatib

Odanacatib was studied by Merck & Co to treat osteoporosis and bone metastasis, which is a cathepsin K inhibitor, an essential cysteine protease enzyme responsible for bone resorption (Fig. **2**) [4, 13]. The isobutyl group in this drug is metabolically unstable due to hydroxylation at this position; however, to solve this problem, the fluorine atom was used to block this position, improving its stability and selectivity. Unfortunately, this drug stoped in phase III clinical trial due to a high risk of stroke.

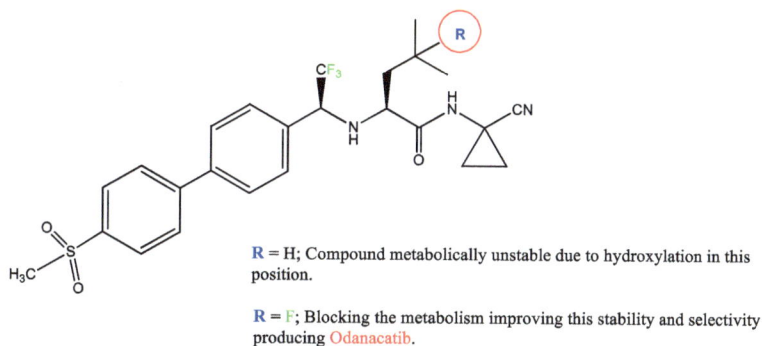

R = H; Compound metabolically unstable due to hydroxylation in this position.

R = F; Blocking the metabolism improving this stability and selectivity producing Odanacatib.

Fig. (2). Structure of the drug Odanacatib.

Celecoxib

The drug Celebrex, also known as celecoxib, is a class of non-steroidal anti-inflammatory drug (NSAID), a COX-2 selective inhibitor approved by the FDA in 1998 (Fig. **3**) [4]. The use of this drug is for different types of pain and inflammation, for example, rheumatoid arthritis, painful menstruation, juvenile rheumatoid arthritis, osteoarthritis, ankylosing spondylitis, and acute pain in adults. Celecoxib is an excellent example of the correct use of fluorine in medicinal chemistry. Its introduction into the aromatic ring increases the half-life in rats in 220 h, making this compound highly stable and difficult to metabolize. Then, a fluorine atom was replaced by a methyl group, which had a much lower half-life of 3.5 h. On the other hand, the introduction of trifluoromethyl group in the heteroaromatic ring increases the potency of this drug [4].

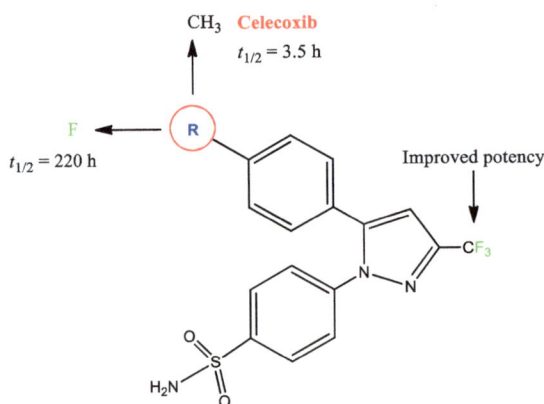

Fig. (3). Development of the drug Celecoxib.

Example

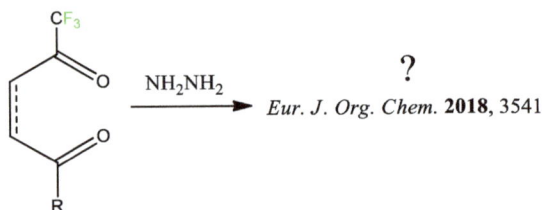

Answer

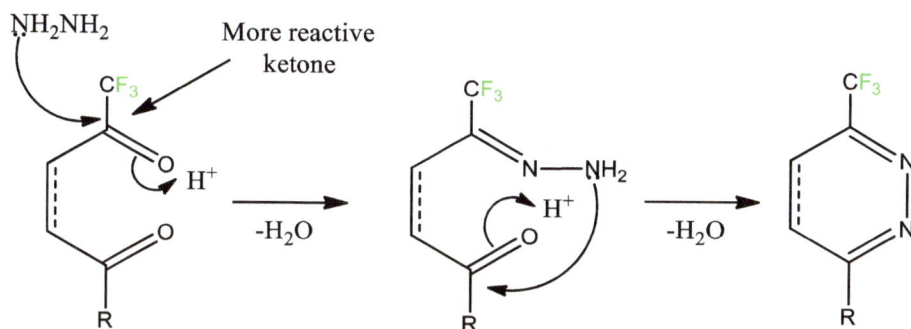

EXERCISES

1) In the exercises **1** to **8** what are the structures of the compounds?

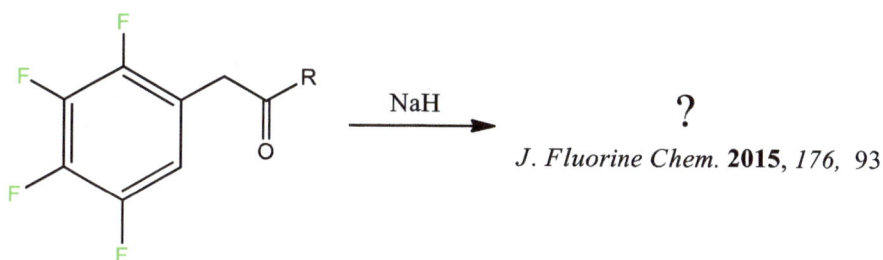

J. Fluorine Chem. **2015**, *176*, 93

2)

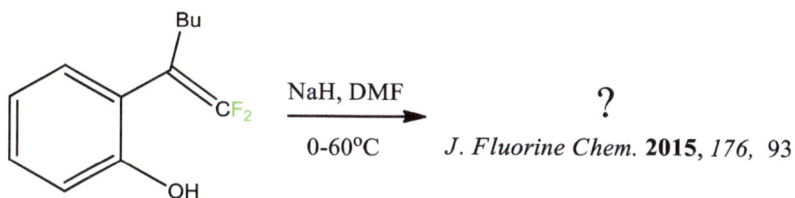

NaH, DMF

0-60°C

?

J. Fluorine Chem. **2015**, *176*, 93

3)

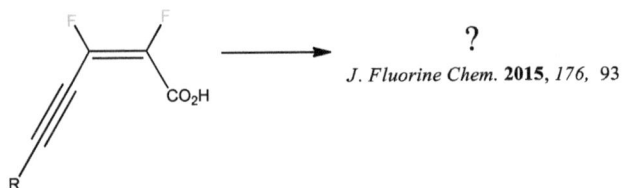

?

J. Fluorine Chem. **2015**, *176*, 93

4)

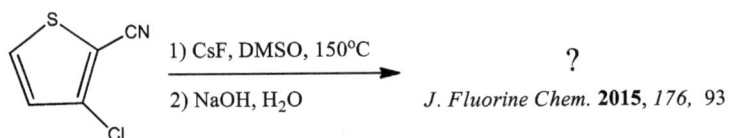

1) CsF, DMSO, 150°C

2) NaOH, H₂O

?

J. Fluorine Chem. **2015**, *176*, 93

5)

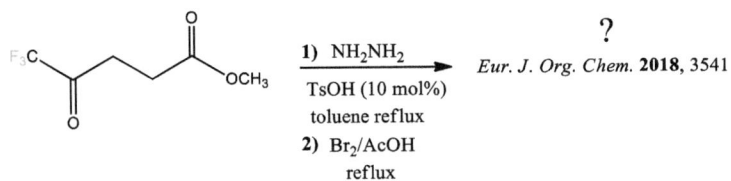

1) NH₂NH₂

TsOH (10 mol%)

toluene reflux

2) Br₂/AcOH

reflux

?

Eur. J. Org. Chem. **2018**, 3541

6)

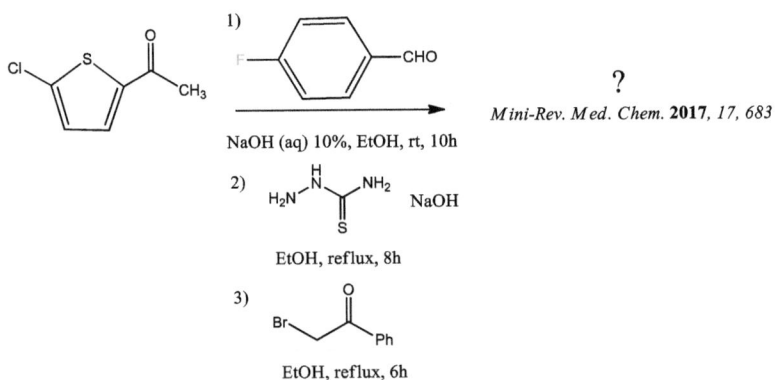

1)

?

Mini-Rev. Med. Chem. **2017**, *17*, 683

NaOH (aq) 10%, EtOH, rt, 10h

2)

NaOH

EtOH, reflux, 8h

3)

EtOH, reflux, 6h

7)

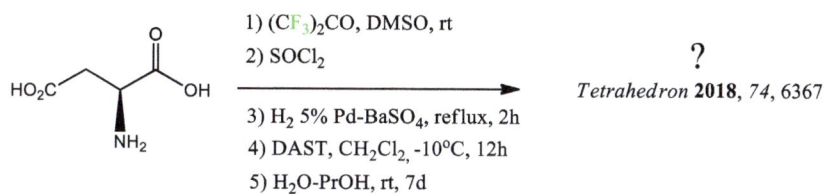

1) (CF$_3$)$_2$CO, DMSO, rt
2) SOCl$_2$

3) H$_2$ 5% Pd-BaSO$_4$, reflux, 2h
4) DAST, CH$_2$Cl$_2$, -10°C, 12h
5) H$_2$O-PrOH, rt, 7d

?

Tetrahedron **2018**, *74*, 6367

8)

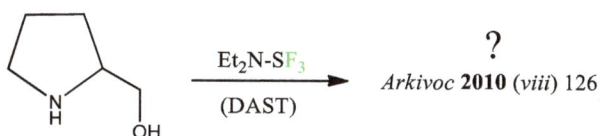

Et$_2$N-SF$_3$

(DAST)

?

Arkivoc **2010** (*viii*) 126

9) In the exercises **9** and **10** propose the mechanism of the reactions.

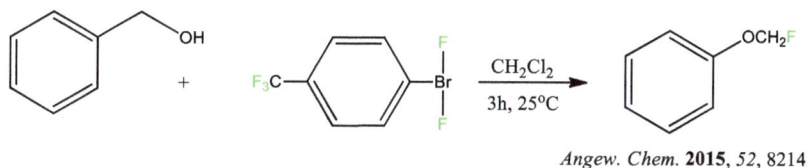

CH$_2$Cl$_2$

3h, 25°C

Angew. Chem. **2015**, *52*, 8214

10)

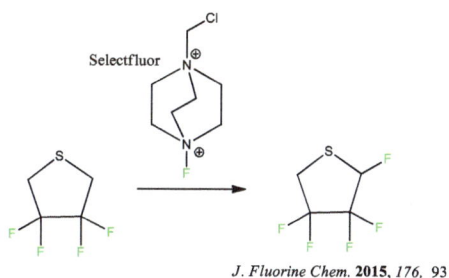

Selectfluor

J. Fluorine Chem. **2015**, *176*, 93

Answers

1)

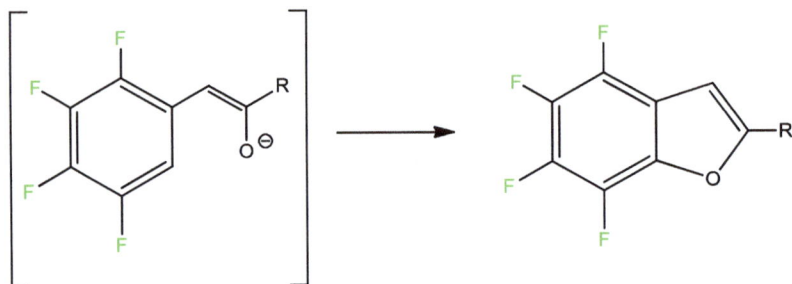

2)

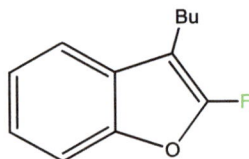

3)

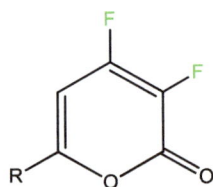

4)

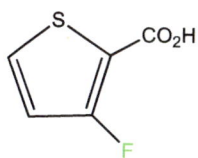

5)

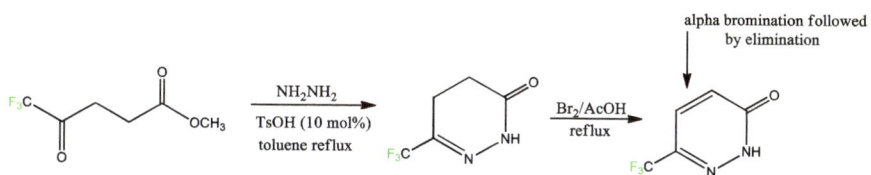

6)

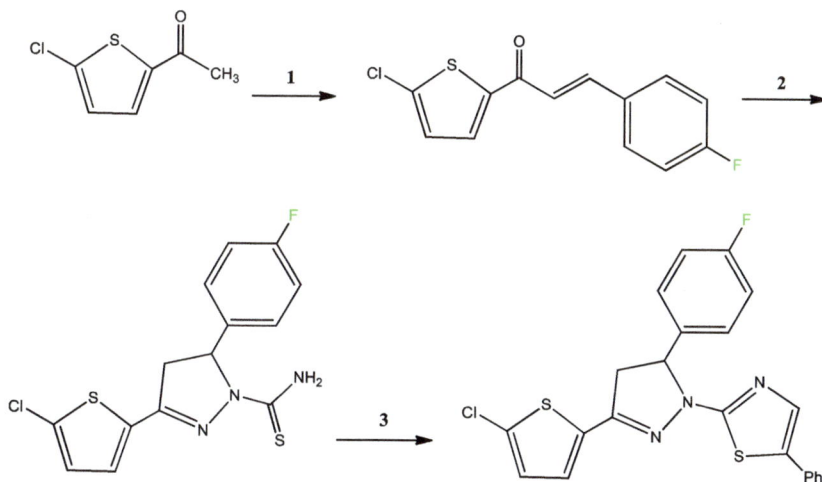

7)

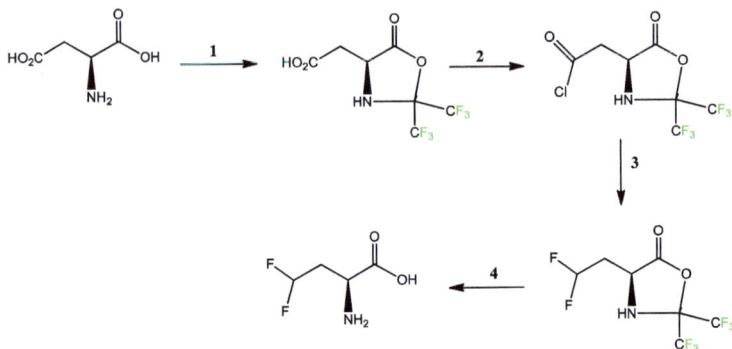

8)

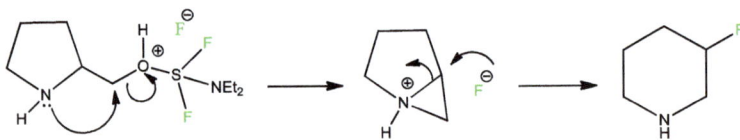

9)

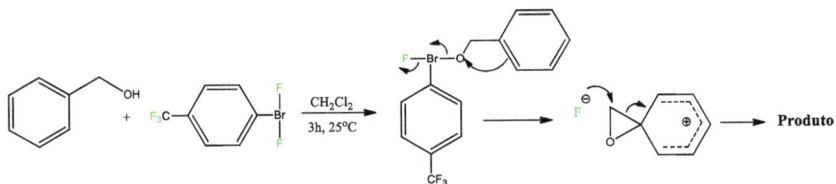

10)

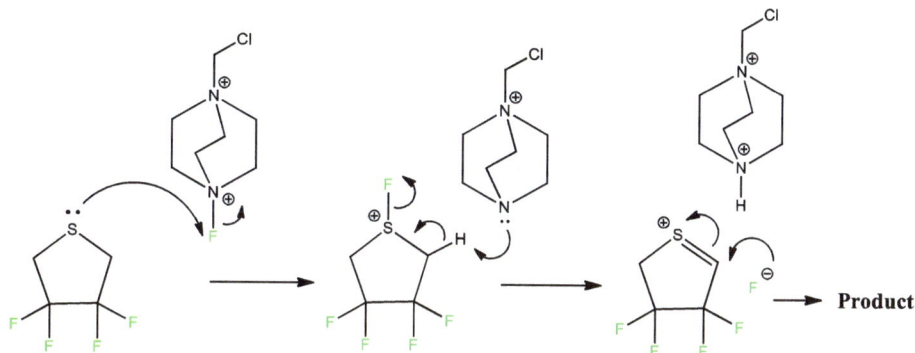

CONCLUSION

Fluorine chemistry is an important area that significantly impacts the drug discovery development, which can be seen by many drugs in the market containing this chemical element. Its presence in a molecule is capable of modifying a series of parameters. Some of these parameters are conformational changes, variations of pKa, the formation of hydrogen bonds, and the increase and decrease in lipophilicity. These characteristics can change both the pharmacodynamics and pharmacokinetic parameters of a bioactive compound. Due to its extreme importance in modern drug discovery, the reader needs to know its properties, chemical reactions and apply it in medicinal chemistry.

REFERENCES

[1]   Wang, B.C.; Wang, L.J.; Jiang, B.; Wang, S.Y.; Wu, N.; Li, X.Q. Shi, Da Y. Application of Fluorine in Drug Design During 2010-2015 Years: A Mini-Review. *Mini Rev. Med. Chem.,* **2017**, *17*(8), 683-692.

[2]   Gupta, S.P. Roles of fluorine in drug design and drug action. *Lett. Drug Des. Discov.,* **2019**, *16*(10), 1089-1109.

[3]   Sharts, C.M. Organic fluorine chemistry. *J. Chem. Educ.,* **1968**, *45*(3), 185.

[4]   Shah, P.; Westwell, A.D. The role of fluorine in medicinal chemistry. *J. Enzyme Inhib. Med. Chem.,* **2007**, *22*(5), 527-540.

[5]   Clark, J.H. Fluoride ion as a base in organic synthesis. *Chem. Rev.,* **1980**, *80*, 429-452.

[6]   Singh, R.P.; Shreeve, J.M. Recent advances in nucleophilic fluorination reactions of organic compounds using deoxofluor and DAST. *Synth,* **2002**, *17*, 2561-2578.

[7]   Milcent, T.; Crousse, B. The main and recent syntheses of the N-CF3 motif. *C. R. Chim.,* **2018**, *21*, 771-781.

[8]   Yin, F.; Wang, Z.; Li, Z.; Li, C. Silver-catalyzed decarboxylative fluorination of aliphatic carboxylic acids in aqueous solution. *J. Am. Chem. Soc.,* **2012**, *134*(25), 10401-10404.

[9]   Becerril, M.R.; Sazepin, C.C.; Leung, J.C.T.; Okbinoglu, T.; Kennepohl, P.; Paquin, J.F.; Sammis,

G.M. Fluorine transfer to alkyl radicals. *J. Am. Chem. Soc.,* **2012**, *134*(9), 4026-4029.

[10] Sazepin, C.C.; Hemelaere, R.; Paquin, J.F.; Sammis, G.M. Recent advances in radical fluorination. *Synth.,* **2015**, *47*(17), 2554-2569.

[11] Suaifan, G.A.R.Y.; Mohammed, A.A.M. Fluoroquinolones structural and medicinal developments (2013–2018): Where are we now? *Bioorg. Med. Chem.,* **2019**, *27*(14), 3005.

[12] Xianhua, P.L.; Xiaojun, L.X.; Qingling, Y.; Wansheng, L.; Weijin, Z.; Qunhui, D.; Fei, L.F. Efficient synthesis of sitagliptin phosphate, a novel DPP-IV inhibitor, *via* a chiral aziridine intermediate. *Tetrahedron Lett.,* **2013**, *54*(50), 6807-6809.

[13] Mei, H.; Han, J.; Klika, K.l.D.; Izawa, K.; Sato, T.; Meanwell, N.A.; Soloshonok, V.A. Applications of fluorine-containing amino acids for drug design. *Eur. J. Med. Chem.,* **2019**, *186*, 111826.

LIST OF JOURNALS USED AND THEIR RESPECTIVE ABBREVIATIONS

Acta Chemica Scandinavica (Acta Chem. Scand.)
Angewandte Chemie, International Edition in English (Angew. Chem. Int. Ed.)
Archiv der Pharmazie (Arch. Pharm.)
Arkivoc (Arkivoc)
Beilstein Journal of Organic Chemistry (Beilstein J. Org. Chem.)
Bioorganic & Medicinal Chemistry (Bioorg. Med. Chem.)
Bioorganic & Medicinal Chemistry Letters (Bioorg. Med. Chem. Lett.)
Bioscience, Biotechnology, and Biochemistry (Biosci. Biotechnol. Biochem.)
Biotechnology and Food Science (Biotechnol. Food. Sci.)
British Journal of Pharmacology (Br. J. Pharmacol.)
Bulletin of the Chemical Society of Japan (Bull. Chem. Soc. Jpn.)
Bulletin of the Korean Chemical Society (Bull. Korean Chem. Soc.)
Catalysis Letters (Catal. Lett.)
Chemical Biology & Drug Design (Chem. Biol. Drug Des.)
Chemical and Pharmaceutical Bulletin (Chem. Pharm. Bull.)
Chemical Communications (Chem. Commun.)
Chemical Research in Toxicology (Chem. Res. Toxicol.)
Chemical Society Reviews (Chem. Soc. Rev.)
Chemistry: A European Journal (Chem. Eur. J.)
Chemistry Letters (Chem. Lett.)
Chinese Chemical Letters (Chin. Chem. Lett.)
Current Topics in Medicinal Chemistry (Curr. Top. Med. Chem.)
Drug Metabolism and Disposition (Drug Metab. Dispos.)
European Journal of Chemistry (Eur. J. Chem.)
European Journal of Medicinal Chemistry (Eur. J. Med. Chem.)
European Journal of Organic Chemistry (Eur. J. Org. Chem.)
Green Chemistry Letters and Reviews (Green Chem. Lett. Rev.)
Helvetica Chimica Acta (Helv. Chim. Acta.)
Heterocycle Communications (Heterocycl. Commun.)
Journal of Agricultural and Food Chemistry (J. Agric. Food Chem.)
Journal of American Chemical Society (J. Am. Chem. Soc.)
The Journal of Antibiotics (J. Antibiot.)
Journal of Antimicrobial and Chemotherapy (J. Antimicrob. Chemother.)
Journal of Carbohydrate Chemistry (J. Carbohydr. Chem.)
Journal of Cell Biology (J. Cell. Biol.)
Journal of Chemical Society, Chemical Communication (J. Chem. Soc., Chem. Commun.)
Journal of Enzyme Inhibition and Medicinal Chemistry (J. Enzyme Inhib. Med. Chem.)
Journal of Heterocyclic Chemistry (J. Heterocycl. Chem.)
Journal of Medicinal Chemistry (J. Med. Chem.)

Journal of Nuclear Medicine (J. Nucl. Med.)
Journal of Organic Chemistry (J. Org. Chem.)
Molecular Pharmacology (Mol. Pharmacol.)
Molecules (não possui abreviação)
Organic & Biomolecular Chemistry (Org. Biol. Chem.)
Organic Communications (Org. Commun.)
Organic Letters (Org. Lett.)
Organic Process Research & Developmen (Org. Process Res. Dev.)
Pharmaceutical Research (Pharmaceut. Res.)
Química Nova (Quim. Nova)
RSC Advances (RSC Adv.)
Russian Chemical Reviews (Russ. Chem. Rev.)
Synlett (não possui abreviação)
Synthesis (não possui abreviação)
Synthetic Communications (Synth. Commun.)
Tetrahedron (não possui abreviação)
Tetrahedron: Asymmetry (Tetrahedron: Asymm.)
Tetrahedron Letters (Tetrahedron Lett.)

LIST OF ABBREVIATIONS

Ac	Acetyl
AD-mix-α	It is a commercially available mixture of the reagents used in the asymmetric dihydroxylation of Sharpless = $K_2OsO_2(OH)_4$ (cat.); K_2CO_3; $K_3Fe(CN)_6$ and (DHQ) 2-PHAL (cat.) = Substance containing two moles of hydroquinine and one mole of phthalazine
AD-mix-β	It is a commercially available mixture of the reagents used in the asymmetric dihydroxylation of Sharpless = $K_2OsO_2(OH)_4$ (cat.); K_2CO_3; $K_3Fe(CN)_6$ and (DHQ) 2-PHAL (cat.) = Substance containing two moles of hydroquinine and one mole of phthalazine
AIBN	Azobisisobutyronitrile
DNA	Deoxyribonucleic acid
LA	Lewis acid
PTSA	*Para*-toluenesulfonic acid
aq.	Aqueous
heat.	Heating
Ar	Aromatic substance
RNA	Ribonucleic acid
TFA	Trifluoroacetic acid
AZT	3'-azido-2 ', 3'-dideoxythymidine
BAIB	Bis(acetoxy)iodobenzene
Bn	Benzyl
Boc	Tert-butoxycarbonyl
Boc₂O	Di-*tert*-butyl dicarbonate
Bu	Butyl
Bz	Benzoyl
C	Catalyst
Cap.	Chapter
cat.	Catalytic
Cbz	Benzyloxycarbonyl
CDI	1,1'-carbonyldiimidazole
MIC	Minimal Inhibitory Concentration
1,5-COD	1,5-cyclooctadiene
conc.	Concentrate
CSA	Camphorsulfonic acid

Cy	Cicloexyl
d	Day or days
DABCO	1,4-diazabicyclo [2.2.2] octane
dba	Dibenzylideneacetone
DBU	1,8-diazabicyclo [5.4.0] undec-7-ene
DCC	Dicyclohexylcarbodiimide
DCM	Dichloromethane
DCU	N, N'-dicyclohexylurea
DDQ	Dicyodichloro-p-benzoquinone
DEAD	Diethyl azodicarboxylate
DET	Diethyl tartrate
DMG	Direct metalation group
DHP	Dihydropyran
(DHQ)₂ and (DHQD)₂PHAL	Substance containing two moles of hydroquinine and one mole of phthalazine.
DIAD	Diisopropyl azodicarboxylate
DIBAL-H	Diisobutylaluminium hydride
DiPAMP	(1S, 2S)-(+)-Bisethane [(2-methoxyphenyl) phenylphosphine]
DIPEA, DIEA or Hunig's Base	N, N-diisopropylethylamine
DIPT	Diisopropyl tartrate
DMA-DMF	N,N-dimethylformamide dimethylacetal
DMAP ou 4-DMAP	4-(dimethylamino)pyridine
DMF	N,N-dimethylformamide
DMP	Dess-Martin Periodinane
2,2-DMP	2,2-dimethoxypropane
DMSO	Dimethyl sulfoxide
dppf	1,1'-bis(diphenylphosphino)ferrocene
E⁺	Electrophile
EDC	1-ethyl-3-(3-dimethylaminopropylcarbodiimide)
EE	Ethoxyethyl
eq.	Equivalent
Et	Ethyl
FDA	Food and Drug Administration

Fmoc	9-Fluorenylmethoxycarbonyl
PG	Protective group
h	Hour ou hours
HATU	2-(1-*H*-7-azabenzotriazol-1-yl)-1,1,3,3-tetramethyluronium hexafluorophosphate
HMDS	Hexamethyldisilazane
HMPA	Hexamethylphosphoramide
hν	Light
HOBt	Hydroxybenzotriazole
HIV	Human immunodeficiency virus
i-**Pr**	Isopropyl
IBX	2-iodoxybenzoic acid
L	Ligant
LDA	*N, N*-diisopropylamide
LDL	Low-density lipoprotein
L-DOPA	L-3,4-dihydroxyphenylalanine
LHA	Lithium aluminum hydride
LHMDS	Lithium hexamethyldisilylamide
m-**CPBA**	*Meta*-chloroperbenzoic acid
MCR	Multicomponente reaction
MEM	2-Methoxyethoxymethyl ether
MO	Microwave
MOM	Methoxymethyl ether
MTM	Methylthiomethyl ether
Ms	Methanesulfonyl (mesylate)
NBS	*N*-bromosuccinimide
NCS	*N*-chlorosuccinimide
NIS	*N*-iodosuccinimide
nm	nanometer
NMP	*N*-methylpyrrolidone
NMO	*N*-methylmorpholine-*N*-oxide
NNRTI	Non-Nucleoside Reverse Transcriptase Inhibitor
Nu	Nucleophile
(O); [O] ou Oxi.	Oxidation
PMB	*Para*-methoxybenzyl
PCC	Pyridinium Chlorochromate

PCy	Cyclohexylphosphine
PDC	Pyridinium dichromate
pg.	Page
Ph	Phenyl
Py	Pyridine
Piv	Pivaloyl
PMHS	Polymethylhydrosiloxane
PPA	Polyphosphoric acid
Pr	Propyl
R^n (n = 0, 1, 2, 3...)	Different substituents
Red-Al	Bis(2-methoxyethoxy)aluminum
RSA	Relationship Structure Activity
rt	Room temperature
NMR	Nuclear magnetic resonance
SEM	β-(trimethylsilyl)ethoxymethyl ether
S$_N$2	Bimolecular nucleophilic substitution
S$_N$i	Substitution Nucleophilic internal
SPhos	2-dicyclohexylphosphino-2,6-dimethoxybiphenyl
TBAI	Tetrabutylammonium iodide
TBDPS	*Terc*-butyldiphenylsilyl
TBAF	Tetrabutylammonium fluoride
TBHP	*Tert*-butyl hydroperoxide
TBS ou TBDMS	*Tert*-butyldimethylsilyl
TEA	Triethylamine
TEMPO	2,2,6,6-tetramethyl-piperidin-1-yl
TES	Triethylsilyl
Tf	Trifluoromethanesulfonate, triflate
THF	Tetrahydrofuran
THP	Tetrahydropyran
TIPS	Triisopropylsilyl
TMEDA	*N,N,N′,N′*-tetramethylethylenediamine
TMS	Trimethylsilyl
TPAP	Tetrapropylammonium perruthenate
Tr	Trityl
Ts	Tosyl

US ou)))	Ultrasound
X	Halogen or a good leaving group
Xantphos	Bis (diphenylphosphino) -9,9-dimethylxanthene
XPhos	2-dicyclohexylphosphino-2',4',6'-triisopropylbipheny

SUBJECT INDEX

A

Acid chloride 11, 48, 61, 88, 89, 90, 91
Acidic condition 89
Acid medium HCl 68, 69
  strong 69
Agrochemicals 209
Alcohol function 87
Alcohols 10, 17, 31, 33, 58, 81, 91, 104, 105,
  212
  benzyl 10, 81
  formed 104, 105
  transform 212
Aldehyde(s) 1, 4, 6, 31, 32, 33, 133, 134, 150,
  176, 177, 181
  carbons 32
  function 32, 33
Aliphatic carboxylic acids 215
Alkylation reaction 91
Alkyl 92, 212, 215
  fluorides 212
  halides 92
  radicals 215
Allyl 105
  bromide 105
  magnesium bromide 105
Amide 88, 89, 90
  concomitant 88
  formation 89, 90
Amine function 88, 90
  chemical 90
Amines 90, 91, 176
  formed 91
  obtaining 90
  primary 176
Amino group regenerates 5
Ammonia 33, 68, 71
  excess 71
Analysis 2, 6, 17
  rational retrosynthetic 17
  retroactive 2
  retrossynthetic 6

Ankylosing spondylitis 219
Antidepressant 23
Antiepileptic rufinamide 199
Anti-HIV drug 3
Anti-inflammatory pifoxime 23
Aromatic 10, 15, 91, 110, 114, 219
  symmetrical 15
Aromatic nucleophilic substitution reaction
  71, 91
Aromatic systems 82
Aromatization 116
Ascorbate sodium 205
Asymmetric epoxidation 69
Atomic fluorine, source of 215

B

Bacterial DNA 216
Barbituric acid 7
  anxiolytic 7
Benzodiazepines 169
Benzophenone 30
Benzyl 2,6-difluorobromide 199
Benzylamine 45
Benzyl 10, 89
  bromide 10
  carbons 89
Benzylic carbon 89
Bicyclic system 73
Bioactive substances 1, 6, 48, 133, 149
  synthetic 48
Biological activity 216
Bone metastasis 218
Breast cancer 150

C

Carbon-carbon bond formation 31, 61
Carbonyl 30, 68, 71, 89
  addition reaction 88
Carboxylic acid 31, 48, 61, 71, 85, 88 89, 90,
  150, 164, 176, 181

T

U

V

W

www.ingramcontent.com/pod-product-compliance
Lightning Source LLC
Chambersburg PA
CBHW080019240326

41598CB00075B/321